GLOBAL RANGELANDS
Progress and Prospects

GLOBAL RANGELANDS
Progress and Prospects

Edited by

A.C. Grice

and

K.C. Hodgkinson
CSIRO, Australia

CABI *Publishing*
in association with
Center for International Forestry Research (CIFOR)

CABI *Publishing* is a division of CAB *International*

CABI Publishing
CAB International
Wallingford
Oxon OX10 8DE
UK

CABI Publishing
10 E 40th Street
Suite 3203
New York, NY 10016
USA

Tel: +44 (0)1491 832111
Fax: +44 (0)1491 833508
E-mail: cabi@cabi.org
Web site: www.cabi-publishing.org

Tel: +1 212 481 7018
Fax: +1 212 686 7993
E-mail: cabi-nao@cabi.org

A catalogue record for this book is available from the British Library, London, UK.

Library of Congress Cataloging-in-Publication Data
Global rangelands: progress and prospects / edited by A.C. Grice and K.C. Hodgkinson.
    p. cm.
    Papers from the 6th International Rangeland Congress, held in Townsville, Australia,
July 17, 1999.
    Includes bibliographical references.
    ISBN 0-85199-523-3 (alk. paper)
      1. Rangelands--Congresses. 2. Range management--Congresses. I. Grice, A. C.
(Anthony C.) II. Hodgkinson, K. C. (Ken C.) III. International Rangeland Congress (6th :
1999 : Townsville, Qld.)

SF84.84 .G56 2002
333.74--dc21

2001052635

ISBN 0 85199 523 3

Typeset by AMA DataSet Ltd, UK.
Printed and bound in the UK by Biddles Ltd, Guildford and King's Lynn.

Contents

Contributors

N. Abel, *CSIRO Sustainable Ecosystems, GPO Box 284, Canberra, ACT 2601, Australia.*

S. Archer, *Department of Rangeland Ecology and Management, Texas A&M University, College Station, TX 77843, USA.*

W. Bayer, *Freelance consultant in pasture management, Rohnsweg 56, D-37085 Göttingen, Germany.*

R. Behnke, *Overseas Development Institute, 111 Westminster Bridge Road, London SE1 7JD, UK.*

R. Blench, *Overseas Development Institute, 111 Westminster Bridge Road, London SE1 7JD, UK.*

A. Bowman, *New South Wales Agriculture, PMB 19, Trangie, NSW 2823, Australia.*

D.M.J.S. Bowman, *Key Centre for Tropical Wildlife Management, Northern Territory University, Darwin, NT 0909, Australia.*

D.D. Briske, *Department of Rangeland Ecology and Management, Texas A&M University, College Station, TX 77843, USA.*

S. Campbell, *Norwood, Blackall, QLD 4472, Australia.*

S. Christiansen, *United States Department of Agriculture Agricultural Research Service, Office of International Research Programmes, Beltsville, Maryland 20706, USA.*

J. Davies, *Faculty of Agriculture and Natural Resources Centre, University of Adelaide, Roseworthy Campus, Roseworthy, SA 5371, Australia.*

S. Díaz, *IMBIV & FCEFyN, Universidad Nacional de Cordoba, C. Correo 495, 5000 Cordoba, Argentina.*

G. Fitzhardinge, *PO Box 35, Mandurama, NSW 2792, Australia.*

B. Foran, *CSIRO Sustainable Ecosystems, PO Box 284, Canberra, ACT 2601, Australia.*

C.D. Freudenberger, *Ethics Consultant, 745 Plymouth Road, Claremont, CA 91711, USA.*

D. Freudenberger, *CSIRO Sustainable Ecosystems, GPO Box 284, Canberra, ACT 2601, Australia.*

M. Gachugu, *ANUTECH, Canberra, ACT 2601, Australia.*

B. Gillam, *Department of Rangeland Management, USDA-Forest Service, Washington, DC 20250, USA.*

A.C. Grice, *CSIRO Sustainable Ecosystems, PMB PO, Aitkenvale, QLD 4810, Australia.*

G. Griffin, *CSIRO Sustainable Ecosystems, PO Box 2111, Alice Springs, NT 0871, Australia.*

R.B. Hacker, *New South Wales Agriculture, PMB 19, Trangie, NSW 2823, Australia.*

T. Hatton, *CSIRO Land and Water, Private Bag 5, Wembley, WA 6014, Australia.*

K.C. Hodgkinson, *CSIRO Sustainable Ecosystems, GPO Box 284, Canberra, ACT 2601, Australia.*

S.M. Howden, *CSIRO Sustainable Ecosystems, GPO Box 284, Canberra, ACT 2601, Australia.*

J. Kotsokoane, *PO Box 1015, Maseru 100, Lesotho, Southern Africa.*

A. Langston, *CSIRO Sustainable Ecosystems, GPO Box 284, Canberra, ACT 2601, Australia.*

M. Lonsdale, *CSIRO Entomology, GPO Box 284, Canberra, ACT 2601, Australia.*

T.J.P. Lynam, *Department of Biological Science, University of Zimbabwe, Box MP167, Mount Pleasant, Harare, Zimbabwe.*

S. Marsden, *CSIRO Sustainable Ecosystems, GPO Box 284, Canberra, ACT 2601, Australia.*

M. Martin, *PO Box 35, Mandurama, NSW 2792, Australia.*

S. McIntyre, *CSIRO Sustainable Ecosystems, 120 Meiers Road, Indooroopilly, QLD 4068, Australia.*

N. Milham, *New South Wales Agriculture, Locked Bag 21, Orange, NSW 2800, Australia.*

D. Mills, *Ennisclare, Gore, QLD 4352, Australia.*

S. Milton, *Faculty of Forestry, Private Bag X1, Matieland 7602, South Africa.*

P.E. Novelly, *Agriculture Western Australia, PO Box 19, Kununurra, WA 6743, Australia.*

W.J. Parton, *Natural Resource Ecology Laboratory, Colorado State University, Fort Collins, CO 80523, USA.*

M. Quirk, *Queensland Department of Primary Industries, PO Box 976, Charters Towers, QLD 4820, Australia.*

E.F. Redente, *Rangeland Ecosystem Science Department, Colorado State University, Fort Collins, CO 80523, USA.*

B. Roberts, *Land Use Studies Centre, University of Southern Queensland, Toowoomba, QLD 4350, Australia.*

G.E. Schuman, *United States Department of Agriculture – ARS, High Plains Grasslands Research Station, 8408 Hildreth R, Cheyenne, WY 82009, USA.*

P. Sloane, *Agricultural and Farming Systems Consultant, Sloane, Cock & King Pty Ltd, Box 1731, North Sydney, NSW 2060, Australia.*

E.L. Smith, *School of Renewable Natural Resources, University of Arizona, Tucson, AZ 85721-0001, USA.*

M. Stafford Smith, *CSIRO Sustainable Ecosystems, PO Box 2111, Alice Springs, NT 0871, Australia.*

D. Tongway, *CSIRO Sustainable Ecosystems, GPO Box 284, Canberra, ACT 2601, Australia.*

B. Walker, *CSIRO Sustainable Ecosystems, GPO Box 284, Canberra, ACT 2601, Australia.*

W. Whitford, *US Environmental Protection Agency, Office of Research & Development, NERL, Environmental Sciences Division, Box 30003, Las Cruces, NM 88003, USA.*

L. Woodhams, *PO Box, Gore, QLD 4352, Australia.*

Preface

The Townsville meeting of global rangeland stakeholders was a significant milestone in the series of International Rangeland Congresses. This sixth Congress drew over 1000 participants from 78 countries including 280 delegates from many developing countries in Africa, Middle East, Asia and South America. It concentrated on rangeland people and their future.

This publication is the result of an attempt to obtain and distil the most important concepts, findings and suggestions for future directions that arose from the various scientific sessions. Summaries are provided by many of the session coordinators. This major distillation covers at least 18 subject areas and represents the most up-to-date description of the state of the art in the global rangeland situation.

The Congress Organizing Committee attempted to include rangeland research, education, extension and development as major building blocks of the Congress. The result was a balanced programme of the science and art of rangeland planning and management: a programme that was based on feedback from our original questionnaire to over 300 international contributors to previous rangeland congresses, including members of the Continuing Committee chaired by Margaret Friedel of Australia.

This publication aims to crystallize out the main streams of thought and discussion subsumed in the two-volume proceedings of the Congress. The hard work and efficiency of our editorial committee of David Freudenberger and David Eldridge produced these volumes. They were ably supported by our production editor, Ann Milligan, who went far beyond the call of duty to produce the full-published proceedings available to delegates on arrival in Townsville on 17 July 1999 – a major achievement for which the organizing committee records its deep gratitude.

It is my sincere hope that this publication will act as an important 'snapshot in time' of the global rangeland position. It can be used as a benchmark which helps us all identify not only what needs to be done, but the priority actions and resources for ensuring a future for rangeland communities and the resources on which they depend. The following chapters draw together the main points and significant outcomes of each Congress session.

It is my earnest hope that this compilation will make a substantial contribution to publication in global rangelands literature and provide a basis for further advances.

Emeritus Professor Brian Roberts
Chairman, Organizing Committee,
VI International Rangeland Congress

Acknowledgements

The editors would like to acknowledge the early contributions made by the Organizing Committee to the development of the scientific programme for the VI International Rangeland Congress. In addition to Ken Hodgkinson and Tony Grice, that Committee consisted of Brian Roberts (Chairman), Andrew Ash, Don Burnside, David Freudenberger, David Eldridge, Ron Hacker and Gordon King. We would also like to express our sincere appreciation to Sonja Slatter for her invaluable contribution in preparing the manuscripts for this publication.

Challenges for Rangeland People

1

Anthony C. Grice and Kenneth C. Hodgkinson

The People of Rangelands

According to Leakey and Lewin (1992) the human species evolved in the rangelands of Africa about 7 million years ago and began spreading around the world around 1 million years ago. Thus people have lived in and exploited lands that we now categorize as rangelands for varying but substantial lengths of time. The ecosystems that typically constitute rangelands – grasslands, savannas, shrublands and woodlands – would have shaped early human development. Rangelands were first used by hunter-gatherer societies that depended on the natural environment for most, if not all, of their needs and this lifestyle prevailed for much of human history.

By around 11,000 years ago, isolated groups of rangeland people began to domesticate animals and plants and to set up subsistence pastoral systems (Diamond, 1998). While depending very strongly on the natural resources of rangelands they increasingly interchanged goods with people developing specific skills and living in non-rangeland areas.

During historical times, rangelands began to be exploited by commercial pastoralists (Walker, 1996) who, like subsistence pastoralists, relied upon domestic livestock to exploit the resources of the rangelands, but who had a much stronger reliance on goods and services from outside the rangelands. Likewise, non-rangeland communities now have very little dependence on the products of commercial rangeland livestock industries. Food and fibre are produced for them from high intensity production systems, often in other parts of the world, and only a few urban people venture into rangelands for recreation.

Hunter gathering, subsistence pastoralism and commercial pastoralism thus represent three human lifestyles that continue to utilize the resources of lands that we now recognize as rangelands. Although all three still exist, there has been a tendency for subsistence pastoralists to replace hunter-gatherers and for commercial pastoralists to replace both as the major (in terms of area utilized) occupiers of rangelands. Indeed, the establishment and development of commercial pastoralism has been the principal means whereby Europeans colonized and then exploited the natural resources of sub-Saharan Africa, Australia, North and South America. The establishment of commercial pastoralism severely weakened most, if not all, indigenous societies.

The Term Rangelands

Explicit recognition of the notion of rangelands occurred only relatively recently. The term 'range' has been used since the 1400s in England to describe extensive areas of land that were either grassed or wooded (*Oxford English Dictionary*, 2000). Early colonists took it to the USA where it came to be associated with extensive, often unenclosed areas of 'natural' lands that were exploited for the grazing of livestock. 'Rangeland' is now an international term but within a country there is a host of substitute terms such as 'wild lands' (USA) and 'outback' (Australia), which mean much the same thing. Rangelands occur in areas of relatively low rainfall or where winters are long and cold. The vegetation is always dominated by natural plant communities rather than by sown pasture.

In general, the human populations of these lands occur at low densities though the large areas of rangeland mean that the total population of humans living in rangelands is significant. On this basis, there are some large differences between so-called 'developed' and 'developing' countries. Human populations of rangelands in developed countries tend to be of lower density than in comparable lands in developing countries (though there may be large urban centres). Furthermore, increasing urbanization at the margins of rangelands is occurring in both the developed and developing parts of the world – a situation that is having major impacts on the use of rangeland resources, especially where people hunt wildlife for recreation.

Currently, the term 'rangelands' focuses on biophysical and land use aspects. This restriction emerged during the 20th century as areas of natural vegetation were increasingly used by people for the production of livestock products from extensive grazing systems. For example, rangelands have been identified as 'uncultivated land that will provide the necessities of life for grazing and browsing animals' (Holechek *et al.*, 1989) and as 'semi-natural ecosystems in which man seeks to obtain a productive output by simply adding domestic stock to a natural landscape' (Harrington *et al.*, 1984). Under commercial pastoralism, the rangelands have provided products that in the main are used by communities of people that live outside the rangelands.

As such, rangelands do not comprise a distinct ecosystem. They have been shaped and defined in large measure by the way humans have used them. Importantly, as described above, human use is always evolving so that during the 20th century the dominant land use over large tracts of rangelands in Australia, USA, Argentina, etc., was commercial pastoralism. In many rangelands, the initial phase of colonization for commercial pastoralism gave way to a phase of consolidation involving increasing refinement of rangeland management. Technologies that helped increase animal production from rangelands were developed and implemented. Fencing, for example, gave greater control over animal movements together with the provision of reliable water supplies for livestock. Typically, degradation of natural resources occurred as a result of this intensification. Re-organization of pastoral businesses and institutions then occurred and the cycle began again (Holling, 1992).

However, there still exists a very broad spectrum of systems for exploiting rangelands – people may have personal preferences that lower the efficiency of resource use and may even degrade the natural resources.

Eventually, commercial pastoralism in rangelands prompted the emergence by the mid-20th century of a formal science to deal with the management of rangelands. Range science departments were established in many universities, professional societies were formed and the publication of range management journals together with national and, later, international rangeland conferences. These developments were centred in, though not exclusive to, the USA. For example, the US-based Society for Range Management was formed in the late 1940s, and its journal, the *Journal of Range Management*, was first published in 1948. Comparable developments took place later in Australia. The Australian Rangeland Society was formed in 1975 and it published the first volume of *The Australian Rangeland Journal* (subsequently *The Rangeland Journal*) in 1976.

The inclusion of both biophysical and land use aspects in the definition of rangelands was integral to the emergence of the science of rangeland management. For example, range management has been defined as 'the science and art of obtaining maximum livestock production from range land consistent with conservation of land resources' (Stoddart and Smith, 1955). The strongest emphases of publications in rangeland journals have related to grazing and grazing management, particularly by domestic livestock, and much of the work reported sought to understand and devise ways of overcoming the biophysical constraints to animal production in rangelands. Research on the social and economic systems of rangelands has, in comparison, received only very minor attention (Walker, 1996).

Four Growing Themes

Against this general background of rangelands as principally constituting a pastoral resource, and of rangeland science aiming at a better understanding of how that resource may be more sustainably and efficiently utilized by domestic livestock, four themes are paramount. These are increasingly challenging traditional definitions of rangelands and are enlarging the scope of rangeland science and management.

The first theme is *multiple uses*. In spite of the strong emphasis on pastoral use, there is a growing recognition that rangelands are used by people in many ways in addition to production of animal products by pastoralism. A number of examples can be given. Mining industries are prominent in many of the world's rangelands. Although utilizing only a very small proportion of the world's rangelands, the mining industries are economically very important, directly and indirectly, for the sustainability of rangeland people and their businesses. Mining does not specifically require a rangeland environment, but where it occurs there may be substantial direct and indirect economic benefits to rangeland people.

Another example is tourism. Today there are pastoral businesses in many countries, both developed and developing, where tourism is an important business component. Sport hunting of various wildlife species, an important tourist activity in the USA and southern Africa, has been culturally and commercially important there and may become increasingly important in other countries.

Provision of ecosystem services for people is another use (United Nations Development Programme *et al.*, 2000). The provision of clean water for human consumption and the natural occurrence of plants, animals and other organisms meet aesthetic and cultural values of rangeland people and are examples of ecosystem services. Large areas of rangelands have also been set aside or reserved specifically to meet conservation, aesthetic and recreational needs.

Finally, many areas of the world's rangelands are home to indigenous peoples. Despite the development of commercial pastoralism, some indigenous human populations continue to use rangeland resources with something resembling their traditional lifestyles and economies.

The theme of multiple uses is not new. For example, its impact on the traditional pastoral use of Australia's rangelands was stressed by Vickery. He wrote:

> Australian rangelands have hitherto been the exclusive domain of the Pastoral Industry, but during the past decade pressures for multiple use embracing recreation, conservation, public works and Aboriginal occupation have emerged, particularly in the Arid Zone, and it is clear that the Pastoralist must be prepared to share the rangeland resource with these land-users on a mutually co-operative basis and without confrontation.

> (Vickery, 1976)

Similarly, Box (1986) criticized the II International Rangeland Congress when he said:

> It did not adequately address rangeland products other than livestock. To focus on commercial pastoralism, a human lifestyle of developed nations, is to further marginalize the people issues of rangelands. More attention to other goods and services will help develop the flexibility needed for the proper use of rangelands.

> (Box, 1986)

In spite of public declaration of the importance of multiple uses of rangelands, subsequent Congresses failed to strongly develop this theme.

The second theme is the *maintenance of the basic resources* upon which all rangeland uses depend. Stoddart and Smith (1955), in their authoritative text on rangeland management, argued that the level of production from rangelands must be commensurate with the long-term maintenance of the resource base. In the language of the late 20th century, this is an issue of sustainability, highlighted by its thematic status during the V International Rangeland Congress held in 1995 (West, 1996). Although self-evident, there are still many rangeland areas that are not routinely monitored in terms of the status and trend in condition of their resources. In many rangelands there is an inadequate understanding of how specific landscapes function and, often, poor commitment by various jurisdictions to monitoring.

The third theme is the importance of *social and economic processes* in resource management. Rangeland science has commonly only dealt with the biophysical constraints of livestock production in rangelands. There have been numerous reminders that factors, other than biophysical ones, also limit management options and the sustainability of rangeland enterprises, industries and communities. It is now recognized that socio-economic factors are often crucial in determining whether particular technical solutions to rangeland problems can be employed in practice (Walker, 1996). Consilience, defined by Wilson (1998) as 'a jumping together of knowledge by the linking of facts and fact-based theory across disciplines to create a common groundwork of explanation', is now required.

The fourth theme concerns *interrelationships*. Rangelands do not exist in isolation from non-rangeland systems. This is true in bio-physical terms because of the natural and anthropogenic interchange of materials and energy between rangelands and non-rangelands. It is also true in socio-economic terms because of the cultural and economic interchanges between rangeland and non-rangeland human communities and societies. Rangelands cannot be managed in isolation from the non-rangeland systems with which they interact. Increasingly, rangelands and the human communities of rangelands are influenced by decisions made principally by communities living outside the rangelands.

This is due to a number of factors. In many countries, increasing urbanization means that smaller proportions of people are directly reliant on the rangelands for their livelihood. On the other hand, the markets for many of the traditional rangeland livestock products have always been outside the rangelands. Further, at least in the so-called 'developed' world, urban populations have a considerable stake in the non-traditional products of rangelands such as water resources and recreational space. Moreover, they express an interest in the condition of the natural resources of the rangelands, often expressed in terms of political pressure. These demands and expectations have the potential to powerfully influence the ways in which rangelands are utilized. Moreover, the influences of non-rangeland peoples upon the uses of rangelands extend across national boundaries as the forces of globalization come to bear on rangeland and urban people alike. Moving geographical boundaries between rangelands and non-rangelands further complicates this. Boundaries shift in response to changing human needs, technologies and climatic circumstances.

The Challenge

The study of rangelands has, since its inception in western USA in the early 20th century, been driven by productivity decline and associated accelerated soil erosion (Young, 2000). It has recognized that rangeland people and their distant 'urban cousins' are somewhat dependent upon the complex, more-or-less natural ecosystems for products of pastoral industries, for recreational opportunities and spiritual enlightenment. Living with this complexity continues to challenge people who live and work in the rangelands. The modern challenges that confront rangeland people are captured in the four growing themes outlined above: multiple uses, sustainability, importance of socio-economic versus biophysical factors, and interactions with non-rangeland systems.

In July 1999, the VI International Rangeland Congress, held in Townsville, Australia, focused on the 'people issues' of the world's rangelands by adopting the theme 'People and rangelands: building the future.' This focus on rangeland people and the need to comprehensively address the four growing themes of the rangelands arose from a challenge given at the previous Congress in Salt Lake City (Walker, 1996). Here, Brian Walker's plenary address stressed that the challenges and opportunities confronting rangelands and their people require far more than technical or scientific solutions. Rangeland science is just one of a number of inputs into sound rangeland management; on its own it is not a major driver of change. Moreover, rangeland management is not simply an issue of pastoral management for natural or semi-natural systems but of utilizing complex ecosystems to meet the needs and expectations of complex human societies. Meeting these challenges requires a diverse range of skills and the input of all stakeholders and may require significant modification of the traditional approach to managing rangelands.

Some of the chapters in this book deal with traditional topics about the use of rangelands for raising livestock. Most emphasize some aspect of the ways in which human communities, societies and institutions influence, or are influenced by, the biophysical properties of rangelands. An opening trilogy focuses on the people of rangelands. 'Future Shocks to People and Rangelands' (Mark Howden, Barney Foran and Roy Behnke) identifies the challenges and issues likely to confront the rangelands and their people in the near future. 'Indigenous People in Rangelands' (Graham Griffin) shows the plight of most indigenous peoples and suggests a way forward for them. 'Rangelands: People, Perceptions and Perspectives' (Denzil Mills, Roger Blench, Bertha Gillam, Mandy Martin, Guy Fitzhardinge, Jocelyn Davies, Simon Campbell and Libby Woodhams) considers the importance of individual and community perceptions of rangelands.

The next ten chapters attempt to link the concerns of rangeland people with the major biophysical components of rangelands. A group of five chapters describes progress and the current status in scientific knowledge about soils, plants and biodiversity. 'Desertification and Soil Processes in Rangelands' (David Tongway and Walter Whitford) and 'Understanding and Managing Rangeland Plant Communities' (Steve Archer and Alison Bowman) concentrate on biophysical aspects of rangelands in relation to soil and plant community processes, respectively, summarizing current understanding of the key processes in each case and identifying where further knowledge would be useful to people. 'Range Management and Plant Functional Types' (Sandra Díaz, David Briske and Sue McIntyre) and 'People and Plant Invasions of the Rangelands' (Mark Lonsdale and Sue Milton) address issues of plant vegetation classification and 'People and Rangeland

Biodiversity' (David Bowman) evaluates the issues connected with nature conservation.

Another group of five chapters, 'Managing Grazing' (Mick Quirk), 'Rehabilitation of Mined Surfaces' (Gerald Schuman and Edward Redente), 'Accounting for Rangeland Resources' (Paul Novelly and Lamar Smith), 'Building on History, Sending Agents into the Future – Rangeland Modelling, Retrospect and Prospect' (Timothy Lynam, Mark Stafford Smith and William Parton) and 'Integrating Management of Land and Water Resources: the Social, Economic and Environmental Consequences of Tree Management in Rangelands' (Tom Hatton), discusses concepts and tools for monitoring rangelands, for modelling processes within rangelands and for understanding and managing grazing, mined surfaces and water resources in rangelands.

A fourth group of chapters discusses issues relating to communication and decision-making processes in rangelands. These include 'Land and Water Management: Lessons from a Project on Desertification in the Middle East' (Scott Christiansen), 'International Perspectives on the Rangelands' (Wolfgang Bayer and Peter Sloane), 'Policies, Planning and Institutions for Sustainable Resource Use: a Participatory Approach' (Nick Abel, Mukii Gachugu, Art Langston, David Freudenberger, Mark Howden and Steve Marsden) and 'Economics and Ecology: Working Together for Better Policy' (Nick Milham).

A fifth group of chapters provides personal perspectives on the future of rangeland people: 'Building the Future: Practical Challenges' (Joe Kotsokoane), 'Rangeland Livelihoods in the 21st Century' (Brian Walker) and 'Building the Future: a Human Development Perspective' (Dean Freudenberger).

Finally, the main insights emerging from this international gathering of rangeland people are synthesized into some key take-home messages in 'Synthesis: New Visions and Prospects for Rangelands' (Ken Hodgkinson, Ron Hacker and Tony Grice). We recognize that this synthesis and the knowledge and insights on which it is based are a snapshot of a changing scene. As such, this book will have a short shelf-life as new issues evolve and old ones fade. Whatever the future, humans in the rangelands will need to be adaptive to survive. This remains the major challenge for rangeland people.

References

Box, T.W. (1986) Perspectives and issues from the Second International Rangeland Congress. In: West, N.E. (ed.) *Rangelands: a Resource Under Siege. Proceedings of the Second International Congress.* Australian Academy of Science, Canberra, pp. 614–616.

Diamond, J. (1998) *Guns, Germs and Steel.* Vintage, London.

Harrington, G.N., Wilson, A.D. and Young, M.D. (1984) *Management of Australia's Rangelands*. CSIRO, Melbourne.

Holechek, J.L., Pieper, R.D. and Herbel, C.H. (1989) *Range Management: Principles and Practices*. Prentice Hall, Englewood Cliffs, New Jersey.

Holling, C.S. (1992) Cross-scale morphology, geometry and dynamics of ecosystems. *Ecological Monographs* 62, 447–502.

Leakey, R. and Lewin, R. (1992) *Origins Reconsidered*. Double Day, New York.

Oxford English Dictionary (online) (2000) Oxford University Press, Oxford.

Stoddart, L.A. and Smith, A.D. (1955) *Range Management*. McGraw-Hill Inc., New York.

United Nations Development Programme, United Nations Environment Programme, World Bank and World Resources Institute (2000) *World Resources 2000–2001: People and Ecosystems: the Fraying Web of Life*. Elsevier Science, Oxford.

Vickery, J. (1976) Guest editorial: Inauguration of the Australian Rangeland Society. *The Australian Rangeland Journal* 1, 5–6.

Walker, B.H. (1996) Having or eating the rangeland cake: a developed world perspective on future options. In: *Rangelands in a Sustainable Biosphere: Proceedings of the Fifth International Rangeland Congress*, Vol. II. Society for Range Management, Denver, Colorado, pp. 22–28.

West, N.E. (1996) *Rangelands in a Sustainable Biosphere: Proceedings of the Fifth International Rangeland Congress*, Vols I & II. Society for Range Management, Denver, Colorado.

Wilson, E.O. (1998) *Consilience: the Unity of Knowledge*. Abacus, London.

Young, J.A. (2000) Range research in the far western United States: the first generation. *Journal of Range Management* 53, 2–11.

Future Shocks to People and Rangelands

2

S. Mark Howden, Barney Foran and Roy Behnke

Introduction

The rangelands of the world have undergone significant changes over the last centuries and decades. More is in store. In some regions there is rapid and fundamental change in the basic socio-economic and political institutions following the removal of state controls on rangelands, precipitating rapid and fundamental alterations to the people and management of the rangelands. In other regions, changes in social, economic and political attitudes and institutions resulting from globalization of markets and other forces are more gradual but still significant. There are changes in the species of plants and animals in many rangeland regions resulting from either deliberate or accidental introductions. Human activities that emit gases like CO_2 and CH_4 are driving atmospheric change that affects the whole globe and may result in climate change at both global and regional levels. Lastly, there is population growth and movement, which occurs in some regions more than others in response to the above and other forces. These different sources of change may interact and are likely to have differential impacts from region to region on the human and biophysical components of the rangelands. These impacts will require considerable adaptation in institutions, technology, management and perhaps expectations of what the rangelands can provide.

We outline some of the possible future changes in the forces operating on and within rangelands, addressing the role that rangeland

©CAB International 2002. Global Rangelands: Progress and Prospects
(eds A.C. Grice and K.C. Hodgkinson)

science may contribute to assessing the impacts of these changes and forming adaptation strategies.

Key Drivers of Change

Early analyses of global change impacts on rangelands focused on climate change impacts on forage and livestock production (e.g. McKeon *et al.*, 1988). Later assessments included the interactive effects of increasing atmospheric levels of carbon dioxide and climate change on production aspects as well as on vegetation distributions (e.g. Allen-Diaz *et al.*, 1996). There was also a growing recognition of the rangelands as contributors to greenhouse emissions as well as the possibility for them to be managed as a sink for carbon. It is only more recently that attempts have been made to integrate broader aspects of ongoing global change (such as the impacts of globalization of markets or major geo-political institutional reform) into assessments of future issues for rangelands.

Foran and Howden (1999) proposed a set of nine major drivers for future change in rangelands: population growth, food security at a national level, globalized trade effects on product prices, institutional capacity for change, energy futures, greenhouse gas emissions, climate change opportunities, urban–rangeland relations and cultural homogenization. To this list needs to be added biological invasions arising from enhanced opportunities for spread of pests and diseases. There are many further issues unlisted. We outline these drivers of future change in the rangelands below and follow with some suggestions of the contribution rangeland science may make in addressing such changes.

Human population growth

World population may grow to more than 10 billion people over the next century. In the developed world, populations may remain largely stable but the less developed world may more than double its population over the next 100 years (Table 2.1). Most of this population growth will occur in urban and more intensively farmed areas. However, growth in local rangeland populations and demand for products from outside the rangelands will place both direct and indirect pressures on the ecosystem integrity of rangelands particularly in developing countries (Wu and Richard, 1999). For example, increased population growth in nations such as Kenya has led to an imbalance of animal numbers and human numbers amongst subsistence pastoralists and the population in general, and a lack of land resources relative to

Table 2.1. Projections of total population (in millions) according to mid-range assumptions of fertility, mortality and migration in the years 2020, 2050 and 2100 (from Lutz, 1996).

	1995	2020	2050	2100
Africa	720	1332	2040	2366
Asia East	1956	2444	2760	2704
Asia West	1445	2228	2995	3136
Europe	808	825	766	624
Latin America	477	693	906	1056
North America	297	359	406	467
Less developed	4451	6541	8554	9137
More developed	1251	1340	1319	1216
World	5702	7879	9874	10350

current and future demand for food and lifestyle (e.g. Prins, 1992). Rangeland institutions can hope to influence local population increase only.

Food security at a national level

To date, the technological basis for world food security appears to be just keeping up with the dual demands of population growth and consumption growth. Per capita grain consumption continues to oscillate around 320 kg, but in 1998 there were only 57 days of consumption in carryover stocks of grain: below the suggested threshold of 70 days (Brown, 1998). In addition, likely future constraints are apparent. Globally, there are continuing trends towards degradation across much of the area used for food production (e.g. Dregne *et al.*, 1991). Whilst some analyses suggest there are still vast areas available for increased productivity (e.g. 550 million ha with some rainfed crop potential which are currently uncropped; Fischer and Heilig, 1997), this broader view of resource availability (enough land with sufficient rainfall) becomes significantly constrained in more focused regional studies. Regions including western Asia, south-central Asia, and northern, eastern, western and southern Africa face future constraints if they wish to feed their populations from land within their own boundaries. For example, nearly all the suitable land in China and all the surface water of India will be required if those two countries are to meet their basic food demands by 2040 (Penning *et al.*, 1997).

Furthermore, the growing affluence of some urban populations in the developing world will demand more meat products whilst rural poverty levels dictate that prospective crop yields may never be attained due to inputs not being adequate to maximize production. These forces, combined with continuing encroachment of cropping on to marginal rangeland areas previously used for grazing, suggest increasing risks of unsustainable resource use. In some locations, this risk is exacerbated by changes in access to resources such as water and dry season grazing which have the capacity to destabilize entire transhumant pastoral systems (de Haan and Gauthier, 1999). Additionally, non-food and bio-energy crops may be more financially competitive than basic food production on the better quality land.

These factors suggest that within the global context described above, rangelands will continue to be minor contributors to food security issues on a regional or a per country basis. Some developed nation rangelands will probably continue to supply animals for fattening in higher rainfall regions, or alternatively for export of live animals to specific niche markets. An optimistic view for rangelands in some developing nations is that they might provide a marginal surplus in livestock products to trade for other food groups and essentials for daily lifestyle. A pessimistic view is that they will just respond to a growing domestic demand for livestock products, which for periods of some decades they might meet (although even now some nations rely on food aid), until resource depletion and climatic events coincide to cause severe degradation of the production system and associated social and economic costs (de Haan and Gauthier, 1999).

Globalized trade and product prices

In 25–50 years' time, the impact of world trade liberalization, globalization of economies and dominance of markets by multinational companies is difficult to foresee. A medium-term view of the next 10 years might predict continued downward price pressure on agricultural and mineral commodities, relieved for short periods by upward price movements brought on by climatic, political or market shocks. Also in the medium term, globalization tends to advantage consumers in richer countries more than workers in poorer countries. The inequitable distribution of world consumption patterns has been maintained or even increased over the last 25 years in spite of the huge changes in the global marketplace (UNDP, 1998).

Rangeland production systems can make only a minor contribution to any world trade effect and thus are driven by it with no ability to control it. To counter this trend, rangeland production systems (in the broadest context) may get leverage from the ideals and rhetoric of sustainable production systems, clean products and regional branding and promotion to provide some market advantage. With the exception of mineral products, the general inability of rangeland ecosystems to provide volume and continuity of supply due to climate variability restrict their economic potential to that of a marginal player in the mainstream commodity markets although they are important in some regional contexts (e.g. live cattle and sheep trade for Australia). The amount of higher rainfall land currently set aside from grazing or other agricultural production in eastern and western Europe and North and South America could, if brought back into production, quickly displace rangeland products if demand increases sufficiently to increase prices.

There is, however, a number of opportunities that might emerge from regionally integrated approaches. The next 50 years will see continued growth in population and affluence in Asia and its gradual emergence as the centre of world food trade (Daviron, 1996). International markets for cereals are stable or declining while markets for meat, fish, fruit and vegetables are expanding. The integration of established dryland industries (pastoralism, tourism, mining) with irrigated systems (e.g. dates, cotton, etc.) may allow some regional rangelands to take advantage of globalization, reversing the trend against declining product prices. However, optimism should be restrained. The volume and continuity constraints mentioned previously, and the doubtful long-term sustainability of arid land irrigation systems given other more economic options for water use and continuing degradation, impose realistic limitations. Furthermore, tourism and investment are highly sensitive to perceived instability in regions and recent experience demonstrates that the magnitude and rapidity of financial flows in the globalized marketplace has the potential to significantly destabilize developing nations. These instabilities and other geo-political changes such as the break-up of the USSR have also reduced the demand for rangeland products such as wool and meat (Kerven and Lunch, 1999).

Institutional capacity for change

In both developed and developing countries, the institutions which are charged with, or assume, management and development responsibilities with respect to rangeland ecosystems and societies have often done badly in the past 100 years, and are often poorly equipped to deal with the next 100 years. For example, degradation crises identified repeatedly over the past 120 years in Australia remain unresolved in spite of dozens of inquiries and the development of substantial institutional structures. The plethora of institutional structures can even generate an 'institutional gridlock' crisis where competing government agencies and their jurisdictional statutes require excessive evaluation of development proposals (Abel, 1999).

In other nations such as China, successive institutional change (from nomadism to state control to privatization of livestock and private control over capital investment) has tended to intensify rangeland use but without recognition of the environmental and socio-economic constraints under which rangelands operate, resulting in marginalization of pastoral communities, increasing inequities and decreasing resource condition (Wu and Richard, 1999). The new sedentary pastoral system has also lost the flexible mechanism that enables effective response to environmental changes. Furthermore, vulnerability to change may be increased by replacing a diverse multi-resource economy with a single-product ranching system with rigid marketing and prices that do not reflect the true cost of production.

In other regions, particularly in the nations which separated from the USSR, there has been a demise of institutions which previously imposed strict controls on land use and management but also provided support through fodder production, financial backing and subsidized inputs (Kerven and Lunch, 1999; Wright and Kerven, 1999). The evolution of new institutional and other arrangements provides an opportunity for these rangeland regions to be more attuned to the needs of a globalized world; however, in the process, there has been considerable human costs, increased inequities and resource degradation (Kerven and Lunch, 1999). The capacity to adjust to future change is uncertain given low levels of capital, degrading infrastructure such as roads and wells, the lack of scientific and technical groups and the uncertain tenure arrangements.

In all these cases, authors have expressed the need for reform to develop locally based institutions which use and enhance indigenous or local knowledge, that recognize cultural, socio-economic and environmental factors, that transfer rights and responsibilities to the local institutions and that develop the meeting places, the language and the processes whereby regional proposals, issues and disputes might be explored and resolved (e.g. Abel, 1999). However, institutions are generally based on precedent not foresight and in a rapidly changing world, there is the likelihood that they will become outdated as markets, cultures and environments change. Thus, an adaptive process needs to be incorporated which can allow for such change. In the past, rangeland science was seen as a key information source for such adaptation. However, an evaluation by Scholes (1999) suggests that classical theoretical, reductionist approaches based on ecological factors only are likely to make limited contributions to such processes. Instead, what is required is a more inclusive, comprehensive framework for analysing rangeland issues which incorporates social, economic and ecological aspects and which provides pathways for implementation of the decisions made by local institutions. These pathways need to link with regional or national governments; however, both experience and theoretical analyses show that responses by governments to local institutions dominated by one group of stake-holders can result in reductions in adaptive capacity rather than increases (Abel, 1999; Janssen *et al.*, 2000). Furthermore, rangeland institutions are usually imposed from, or dominated by, forces from outside the rangelands (e.g. Abel, 1999; Wu and Richard, 1999). Thus there is little feedback to, or control of, those external institutional forces to initiate such change (Ebohon *et al.*, 1997; Savich, 1998).

Energy futures

The development of easily accessed and deliverable fossil fuel supplies has underpinned growth and development in the 20th century. For rangeland industries this has meant easy and relatively cheap access to transport, communications and advanced technology and infra-structure. These are all critical in overcoming the disadvantages of distance and location that are inherent in rangelands. This era in range-land development may change in the next 25 years as supplies of conventional oil are depleted. Petroleum analysts such as Campbell (1998) caution that by 2015 the lack of major new oil finds and an expanding rate of consumption may rapidly deplete the world's conventional oil reserves.

Alternative viewpoints to those of Campbell abound. These relate to how depleting conventional oil supplies will encourage the search for more oil and gas, the extraction of unconventional supplies (oil from coal or oil shale, gas from methyl hydrates on the ocean floor), the introduction of a biomass-based fuel cycle (methanol, ethanol, hydrogen) or rapid technological change leading to a four- to tenfold reduction in fuel consumption for the delivery of the same service. In most cases, there are likely to be increases in either the price of the alternative fuel or in the new, more efficient technologies. This will increase costs for rangelands, further reducing already declining terms of trade.

Greenhouse gas emissions

Whatever the approximation of the energy scenarios described above, there seems little doubt that atmospheric concentrations of carbon dioxide will double pre-industrial levels (280 p.p.m.) by the end of this century. This will produce a number of direct and indirect bio-physical effects in rangeland ecosystems but perhaps more importantly may give rise to a profound change in the structure and function of world, national and regional governance, institutions and industrial metabolism arising from carbon trading. These changes would be in response to a recognition that even the least emission-intensive of the future scenarios developed by the Intergovernmental Panel on Climate Change (Nakicenovic and Swart, 2000) suggests that atmospheric concentration of carbon dioxide will rise from 364 p.p.m. currently to 600 p.p.m. by 2100 with a corresponding rise in average global temperature of about 1.5° to 4°C and other climatic changes. Climatic changes may generate additional risks to food, fibre and forestry production, human health, infrastructure and the natural environment (Watson *et al.*, 1997).

The aim of the Framework Convention on Climate Change is to limit these risks and the Kyoto Protocol is the first step in limiting emissions of greenhouse gases. The Protocol commits most developed nations to reducing emissions compared with 1990 levels as well as fostering technology transfer to developing nations through various activities. In line with the main theme of marginalization in this chapter we anticipate that rangeland areas will accept policies rather than set them. However, the combination of lower population densities than other regions, larger areas, limited ecosystem productivity and fewer vested interests might provide unanticipated opportunities in a global carbon market.

It is too early to be definitive about the implications for emission trading for rangelands, but it is likely that there may be both opportunities and costs. Opportunities arise from the possibility of storing carbon in managed vegetation (2.5–25 t C ha^{-1}, Glenn *et al.*, 1993; 20–30 t C ha^{-1}, Moore *et al.*, 1997) and in rehabilitation of degraded soils (8 t C ha^{-1}, Ash *et al.*, 1996) for purchase as emission offsets by countries and industries that emit carbon dioxide from the use of fossil energy. Such arrangements may offer a 30–50-year adaptation period until the carbon pool on a particular area of land is filled to capacity, or a longer period if successive parcels of land are taken up and managed for carbon accumulation. Disadvantages arise from the greater methane emissions per unit product and per unit economic return for rangeland livestock when compared with those from livestock in more mesic regions stemming from greater emissions per unit feed intake and lower rates of productivity (Howden and Reyenga, 1999; Kurihara *et al.*, 1999) and the potentially greater costs of fossil fuel-based inputs.

Climate and atmospheric change impacts

In addition to possible impacts on rangelands relating to emissions trading, there may be impacts arising directly from increased levels of CO_2 in the atmosphere and from the associated climate changes. The impact of these global changes will be experienced differentially by latitude, nation and region (Watson *et al.*, 1997). Yet again the rangelands will be affected by the global changes and may have comparatively fewer opportunities for adaptation, although some are noted below.

Increasing CO_2 and temperature and rainfall changes might allow some rangelands in developed nations to become more productive in plant, animal and financial terms (e.g. Campbell *et al.*, 1997; Hall *et al.*, 1998; Howden *et al.*, 1999b) provided there is not a large decrease in rainfall or an increase in El Niño-like climate events. However, increases in animal production may be partly offset by greater frequencies of thermal stress on the grazing animals (Howden *et al.*, 1999a). Land use options in these rangelands may also be expanded to include more intensive uses such as cropping in the wetter margins where soil types permit (e.g. Howden *et al.*, 1999b; Reyenga *et al.*, 1999). Trade effects induced by disruption or enhancement of production systems in other countries may have significant, currently uncertain impacts (Parry *et al.*, 1999).

The effects on rangelands in Asia will be at best neutral and perhaps submerged by biophysical issues such as hydrological changes in the Himalayas and the impact of population growth over the entire region. Assessments that crop yields will generally decrease with climate change notwithstanding increased CO_2 effects (McLean *et al.*, 1997) suggest that increased pressure will be exerted on the rangelands for food production as is already happening now (Wu and Richard, 1999). This will be exacerbated by possible reductions in grazing areas and productivity from expansion of timberlines and increases in aridity (Wu, 1999). Increase in disease risk may have significant implications for human populations (McLean *et al.*, 1997).

The impacts of temperature rise and rainfall change in African rangelands could be most severe because of the challenged nature of the rangeland resources currently, and the prospect of further population increase, institutional decay and resource degradation (Zinyowera *et al.*, 1997). However, increased atmospheric CO_2 concentrations may partly offset the impacts of periodic droughts, making grasslands more resilient to climate variability and human influences in those areas where soil nutrients are adequate (e.g. Campbell *et al.*, 1997). A global land-use change study (IIASA, 1999) found that the interzone area between rangelands and croplands in western Kenya could be advantaged under rising temperature regimes giving larger areas of cultivation potential and allowing higher altitude areas to be brought under cultivation. In this example the issues described in previous rangeland drivers, particularly population growth and institutional decay, could undercut the biophysical assessment.

Urban–rangeland relations

Since 1950, the number of people living in the world's urban areas has jumped from 750 million to 2.64 billion people or 46% of the total (up from 30% in 1950) (Brown, 1998). Every week more than 1 million people are added to urban centres, and by the year 2000 nearly half of the world's poorest people, some 420 million, were urban. The investment and management challenge required to meet basic habitation requirements in the developing world, and to remake the developed world's cities into pleasant liveable places, will dwarf the challenges of rangeland management. Thus rangelands will be driven by the side effect of the main influences rather than participating in them.

The rangeland implications of this urban growth is that rangelands have become, and will continue to be, distanced from the real affairs of the majority of people and their political decisions. As a percentage of total populations, probable rangeland populations account for 2%, 15% and 3% for developed nation, African and South Asian regions respectively (UNDP, 1998). In the past century the economic importance of rangelands, their myths and their legends have served them well in terms of political activity, international visibility and investment decisions. The 21st century is likely to herald the start of the megacity millennia, where large concentrations of urban people are seen as the central point of economic growth and the investment needed to sustain it. The relative importance of rangelands might only be maintained if they supply services and products central to the maintenance and survival of those urban concentrations. This supply may be dependent on the other drivers discussed here.

Cultural homogenization

The rangelands of the world still harbour original, relatively intact and stunningly different human cultures. African pastoral tribes such as the Masai and the Boran, the Aboriginal peoples of Australia and the Indian tribes of North America are examples. However, homogenization of the cultures of the rangelands, akin to that already occurring in the world's cities, may occur from a number of forces. These include uniformity of production methods to meet mainstream commodity standards at the expense of traditional methods and livestock types, constraining of management and investment activities to stereotypes that meet the perceived needs of global financial markets and the expectations of tourists for constant levels of facilities and access. The increased exposure of rangeland peoples to external media may also lead to greater homogeneity of expectations in terms of lifestyles and material goals. Furthermore, the communications revolution and the rapid delivery of information through the Internet may have large impacts by itself. For example, what were once vastly different lands and peoples, the remote rangelands are now more and more simply market places or tourism destinations that are becoming recognizable, but similar, the world over. Camel rides, humped cattle, romantic tribesmen and wilderness are now accessed through the travel agent or the Internet provider. Once a rangeland has been experienced, tasted and sampled, the global consumer may move at will to the next option on the global menu. We doubt if rangelands can resist these pressures although the theme of marginalization may help.

Globalization forces may also engender counter-currents that lead to cultural differentiation rather than homogenization. For example, currency and political instabilities arising from globalization of financial markets are currently resulting in often violent re-establishments of ethnic and regional culture. Internet communications are providing some Aboriginal groups in arid Australia the means to re-establish cultural linkages and to be more politically effective in gaining land rights and access to health and education. Similarly, the Internet is a potent means of differentiation of tourism markets in ways that can maintain local cultural integrity if managed by the local inhabitants.

There is thus a tension between the potentially destructive and constructive elements of globalization in regard to the development and preservation of rangelands cultures. Rejecting homogenization requires cultural visions that are developed and maintained at multiple scales (i.e. local, regional, national and international).

Invasive Species

Homogenization is also occurring in the flora and fauna of rangelands through both the deliberate and accidental introduction of non-native plants, animals and diseases (Huenneke, 1999). Deliberate introductions arise through the human drive to introduce species for enhanced food and fibre production, resource conservation and other uses. Accidental introductions arise through seed lots, feedstuffs, transport along roads and other transport routes. The likelihood of accidental introduction seems to be increasing because of reduced quarantine standards due to governmental cost-cutting and changing trade regulations, increased global mobility, increased road use and increased disturbance including that arising from tourism and leisure activities. Establishment could be being enhanced by increased nitrogen availability from atmospheric deposition from urban and industrial centres and from increased use of nitrogen-fixing legumes in pastures. These factors are generally expected to trend upwards leading to increased potential for invasions (Huenneke, 1999). In addition, increases in CO_2 concentrations, increased climate-related disturbance (droughts, floods, storms, heavy rainfall) from climate change and changes in geographical ranges of pests and diseases with climate change may also increase problems arising from such invasive species. Improved cost–benefit analyses are needed to evaluate this issue (Mack, 1999) along with perhaps revised ideas of the desirable species mix of rangeland communities.

The Roles of Science in Adapting to Future Shocks

Extrapolation of these drivers over the next 200 years suggests that existing downward trends in ecological, social and economic status are likely to continue and rangelands are likely to become more marginalized in world and national affairs than they have been in the past and will suffer continuing and substantial shocks. An additional aspect of this marginalization is that in most cases there is a lack of feedback from the rangelands into the causal agents of these drivers thus providing little capacity for modifying them. Consequently, Foran and Howden (1999) suggest that a key strategy is establishing adaptive mechanisms to such changes. Components of such a strategy are to start to redesign local institutions, to better integrate science with social and economic considerations and to think laterally and creatively about how to best use the human and other resources in rangelands. Rangeland science can contribute to the development and implementation of these strategic tasks, but in many cases this will require both a substantial change in *modus operandi* towards approaches which integrate biophysical, cultural and economic goals as well as a commitment to continuing adaptation.

If historical institutional arrangements have failed to maintain the function and resilience of rangeland ecosystems and societies, increasing marginalization can only increase the degree of institutional failure. Measures to construct more appropriate institutions will include some of the following:

- Intervention by, and enthusiasm from, local communities to take control of their own development destinies and to use science in developing and implementing their visions.

- Development of appropriate fora and methods of information exchange (stories, language, data, experience, culture, structured interactions, information technology) which enable equitable participation by all stakeholders and more effective involvement of science.
- Foster the rationale, skills and capacity at a regional level to interpret the local implications of externally imposed legislation or other changes, and provide a certification venue for regional development proposals which includes scientific assessment.
- Devise theory and design for new institutions which are sensitive to the dynamics of societies and ecosystems but which transfer learning and adaptation so that local ownership and action of scientific information is maintained.

The new science approaches in rangelands veer inevitably towards maintaining resource function, diminishing the effects of marginalization, the development of adequate lifestyle and living infrastructure and redressing the inequities of poverty, education and future opportunity. The science itself must be driven by the challenge of integrating biophysical, social and economic factors and by specific issues such as future energy options, climate and atmospheric change and carbon storage opportunities. Some emergent themes are as follows:

- Development of systems and theory which allow regional institutions and their stakeholders to integrate biophysical, social and economic dynamics and to compare alternative development strategies over timeframes spanning human generations (25–50 years) (e.g. Abel, 1999).
- Developing new modes of habitation and service provision which embody low levels of energy and materials, have low fossil energy running requirements and which maximize the local use of labour for construction and maintenance.
- Modelling the material flows needed to maintain a rangeland regional economy and using industrial ecology concepts to design alternative regional structures and functions which attempt to 'close the loop' on water, material and nutrient flows.
- Designing ways of living in rangelands that are less dependent on transport in and transport out to reduce reliance on possibly erratic supplies of fossil fuels or expensive alternatives. For example, use information technology to substitute for material movement where possible.

Attempting to overcome the limitations imposed by distance, geomorphology and climate variability has been the goal of rangeland science for the past 50 years or more. However, these limitations of rangelands might have produced somewhat fortuitously a number of comparative advantages. In spite of past and current degradation, many rangelands are still essentially natural and lack the pollution problems in so much of the world's croplands and highly modified peri-urban areas. For example, grazing industries can target higher quality markets that demand and pay for 'naturalness' and for features such as freedom from pesticide and herbicide residues and from diseases such as BSE (bovine spongiform encephalopathy). Science has a role in identifying culturally and environmentally appropriate production systems to meet these market needs. Similarly, increasing woody vegetation, until now the *bête noire* of many woody weed ecologists, may provide a sustaining income for one to two generations of pastoralists as carbon trading attempts to help urban and industrial societies make the transition to a new energy economy. Science has a role in quantifying the trade-offs between different land uses and in developing effective measurement and monitoring techniques. While distance and erratic productivity were seen as limitations to livestock productivity in the past, the space and naturalness that define rangelands can now be promoted as their greatest cultural asset in a world where more than 50% of all people will live in densely populated urban areas. For example, instant information transfer can allow erratic production of uniquely rangeland 'clean and green' products to be used as a marketing edge to urban consumers seeking experiences different from those with commodities that are constantly available. There are possibilities for the use of seasonal forecasting to enhance this marketing capacity.

Conclusion

Throughout this chapter the emphasis has been on rangeland science needing to be better integrated with the cultural, institutional, business and environmental concerns of the rangeland stakeholders. This will require a significant departure from many of the past practices for rangeland science but will ensure that it remains relevant in an uncertain future.

References

Abel, N.O.J. (1999) Preparing for the future: rangelands as complex adaptive systems. In: Eldridge, D. and Freudenberger, D. (eds) *People of the Rangelands. Building the Future. Proceedings of the VI International Rangeland Congress.* VI International Rangeland Congress Inc., Townsville, Australia, pp. 21–30.

Allen-Diaz, B., Chapin, F.S., Diaz, S., Howden, S.M., Puidefabregas, J., Stafford-Smith, M., Benning, T., Bryant, F., Campbell, B., du Toit, J., Galvin, K.,

Holland, E., Joyce, L., Knapp, A.K., Matson, P., Miller, R., Ojima, D., Polley, H.W., Seastedt, T., Suarez, A. and Svecar, T. (1996) Rangelands in a changing climate: impacts, adaptation and mitigation. In: *IPCC Second Assessment Report*. WMO/UNEP, Geneva, pp. 131–158.

Ash, A.J., Howden, S.M. and McIvor, J.G. (1996) Improved rangeland management and its implications for carbon sequestration. In: *Proceedings of the V International Rangeland Congress*, Utah, July 1995, pp. 19–20.

Brown, L.R. (1998) *Vital Signs 1998*. W.W. Norton and Company, New York.

Campbell, B.D., Stafford Smith, D.M. and McKeon, G.M. (1997) Elevated CO_2 and water supply interactions in grasslands: a pasture and rangelands management perspective. *Global Change Biology* 3, 177–87.

Campbell, C.J. (1998) *The Coming Oil Crisis*. Multi-Science Publishing and Petro Consultants SA, Brentwood, UK.

Daviron, B. (1996) Some significant facts concerning the recent dynamics of the trade in food products. *Economie Rurale* 234/235, 10–16.

de Haan, C. and Gauthier, J. (1999) The management of shocks to people and rangelands. In: Eldridge, D. and Freudenberger, D. (eds) *People of the Rangelands. Building the Future. Proceedings of the VI International Rangeland Congress*. VI International Rangeland Congress Inc., Townsville, Australia, pp. 1060–1062.

Dregne, H.E., Kassas, M. and Rozanor, B. (1991) A new assessment of the status of desertification. *Desertification Control Bulletin* 20, 6–18.

Ebohon, O.J., Field, B.G. and Ford, R. (1997) Institutional deficiencies and capacity building constraints – the dilemma for environmentally sustainable development in Africa. *International Journal of Sustainable Development and World Ecology* 4, 204–213.

Fischer, G. and Heilig, G.K. (1997) Population momentum and the demand on land and water resources. *Philosophical Transactions of the Royal Society of London* 352, 869–889.

Foran, B. and Howden, S.M. (1999) Nine global drivers of rangeland change. In: Eldridge, D. and Freudenberger, D. (eds) *People of the Rangelands. Building the Future. Proceedings of the VI International Rangeland Congress*. VI International Rangeland Congress Inc., Townsville, Australia, pp. 7–13.

Glenn, E., Squires, V., Olsen, M. and Frye, R. (1993) Potential for carbon sequestration in the drylands. *Water, Air and Soil Pollution* 70, 341–355.

Hall, W.B., McKeon, G.M., Carter, J.O., Day, K.A., Howden, S.M., Scanlan, J.C., Johnston, P.W. and Burrows, W.H. (1998) Climate change in Queensland's grazing lands. II. An assessment of the impact on animal production from native pastures. *Rangeland Journal* 20, 151–176.

Howden, S.M. and Reyenga, P.J. (1999) Methane emissions from Australian livestock: implications of the Kyoto Protocol. *Australian Journal of Agricultural Research* 50, 1285–1291.

Howden, S.M., Hall, W.B. and Bruget, D. (1999a) Heat stress and beef cattle in Australian rangelands: recent trends and climate change. In: Eldridge, D. and Freudenberger, D. (eds) *People of the Rangelands. Building the Future. Proceedings of the VI International Rangeland Congress*. VI International Rangeland Congress Inc., Townsville, Australia, pp. 43–45.

Howden, S.M., McKeon, G.M., Walker, L., Carter, J.O., Conroy, J.P., Day, K.A., Hall, W.B., Ash, A.J. and Ghannoum, O. (1999b) Global change impacts on native pastures in south-east Queensland, Australia. *Environmental Modelling and Software* 14, 307–316.

Huenneke, L.F. (1999) A helping hand: facilitation of plant invasion by human activities. In: Eldridge, D. and Freudenberger, D. (eds) *People of the Rangelands. Building the Future. Proceedings of the VI International Rangeland Congress.* VI International Rangeland Congress Inc., Townsville, Australia, pp. 562–566.

IIASA (1999) Kenyan land use change study. World wide web address: http://www.iiasa.ac.at/Research/LUC/WP/wp96071_sum.html

Janssen, M.A., Walker, B., Langridge, J. and Abel, N. (2000) An adaptive agent model for analysing co-evolution of management and policies in a complex rangeland system. *Ecological Modelling* 131, 249–268.

Kerven, C. and Lunch, C. (1999) The future of pastoralism in Central Asia. In: Eldridge, D. and Freudenberger, D. (eds) *People of the Rangelands. Building the Future. Proceedings of the VI International Rangeland Congress.* VI International Rangeland Congress Inc., Townsville, Australia, pp. 1042–1048.

Kurihara, M., Magner, T., Hunter, R.A. and McCrabb, G.J. (1999) Methane production and energy partition of cattle in the tropics. *British Journal of Nutrition* 81, 263–272.

Lutz, W. (1996) *The Future Population of the World: What Can We Assume Today?* Earthscan Publications, London (revised and updated edition).

Mack, R.N. (1999) The motivation for importing potentially invasive plant species: a primal urge? In: Eldridge, D. and Freudenberger, D. (eds) *People of the Rangelands. Building the Future. Proceedings of the VI International Rangeland Congress.* VI International Rangeland Congress Inc., Townsville, Australia, pp. 557–562.

McKeon, G.M., Howden, S.M., Silburn, D.M., Carter, J.O., Clewett, J.F., Hammer, G.L., Johnstone, P.W., Lloyd, P.L., Mott, J.J., Walker, B., Weston, E.J. and Wilcocks, J.R. (1988) The effect of climatic change on crop and pastoral production in Queensland. In: Pearman, G.I. (ed.) *Greenhouse. Planning for Climate Change.* CSIRO, pp. 546–563.

McLean, R.F., Sinha, S.K., Mirza, M.Q. and Lal, M. (1997) Tropical Asia. In: Watson, R.T., Zinyowera, M.C., Moss, R.H. and Dokken, D.J. (eds) *The Regional Impacts of Climate Change: an Assessment of Vulnerability: Summary for Policy-makers.* A special report of IPCC Working Group II. Cambridge University Press, Cambridge, UK, pp. 381–407.

Moore, J.L., Howden, S.M., McKeon, G.M., Carter, J.O. and Scanlan, J.C. (1997) A method to evaluate greenhouse gas emissions from sheep grazed rangelands in south west Queensland. *Modsim '97 International Congress on Modelling and Simulation Proceedings*, 8–11 December, University of Tasmania, Hobart (McDonald, D.A. and McAleer, M. eds). Modelling and Simulation Society of Australia, Canberra, pp. 137–142.

Nakicenovic, N. and Swart, R. (2000) *Special Report on Emissions Scenarios.* Intergovernmental Panel on Climate Change. Cambridge University Press, Cambridge, UK.

Parry, M., Rosenzweig, C., Iglesias, A., Fischer, G. and Livermore, M. (1999) Climate change and world food security: a new assessment. *Global Environmental Change* 9, 51–57 (supplement).

Penning, F.W.T., de Vries, R., Rabbinge, R. and Groot, J.J.R. (1997) Potential and attainable food production and food security in different regions. *Philosophical Transactions of the Royal Society of London* 352, 917–928.

Prins, H.H.T. (1992) The pastoral road to extinction: competition between wildlife and traditional pastoralists in East Africa. *Environmental Conservation* 19, 117–123.

Reyenga, P.J., Howden, S.M., Meinke, H. and Hall, W.B. (1999) Global change impacts on wheat production along an environmental gradient in South Australia. In: *Modsim '99 International Congress on Modelling and Simulation Proceedings*, 6–9 December, Hamilton, New Zealand, pp. 753–758.

Savich, H.V. (1998) Global challenge and institutional capacity – or how can we refit local administration for the next century. *Administration and Society* 30, 248–273.

Scholes, R.J. (1999) Does theory make a difference? In: Eldridge, D. and Freudenberger, D. (eds) *People of the Rangelands. Building the Future. Proceedings of the VI International Rangeland Congress*. VI International Rangeland Congress Inc., Townsville, Australia, pp. 31–34.

UNDP (1998) *Human Development Report 1998*. Oxford University Press, Oxford, UK, 228 pp.

Watson, R.T., Zinyowera, M.C., Moss, R.H. and Dokken, D.J. (1997) *The Regional Impacts of Climate Change: an Assessment of Vulnerability: Summary for Policy-makers*. A special report of IPCC Working Group II, November 1997, 22 pp.

Wright, I.A. and Kerven, C. (1999) The role of winter fodder in livestock systems on rangelands in Kazakstan. In: Eldridge, D. and Freudenberger, D. (eds) *People of the Rangelands. Building the Future. Proceedings of the VI International Rangeland Congress*. VI International Rangeland Congress Inc., Townsville, Australia, pp. 47–48.

Wu, N. (1999) Vegetation patterns in West Sichuan, China and man's impacts on its dynamics. In: Miehe, G. (ed.) *Environmental Changes in High Asia: Proceedings of an International Symposium at Marburg, Germany, 28–31 May 1997*. Marburg University, Marburger Geographische Schriften.

Wu, N. and Richard, C. (1999) The privatization process of rangeland and its impacts on the pastoral dynamics in the Hindu-Kush Himalaya: the case of western Sichuan, China. In: Eldridge, D. and Freudenberger, D. (eds) *People of the Rangelands. Building the Future. Proceedings of the VI International Rangeland Congress*. VI International Rangeland Congress Inc., Townsville, Australia, pp. 14–21.

Zinyowera, M.C., Jallow, B.P., Maya, R.S. and Okoth-Ogendo, H.W.O. (1997) Africa. In: Watson, R.T., Zinyowera, M.C., Moss, R.H. and Dokken, D.J. (eds) *The Regional Impacts of Climate Change: an Assessment of Vulnerability: Summary for Policy-makers*. A special report of IPCC Working Group II. Cambridge University Press, Cambridge, UK, pp. 29–84.

Indigenous People in Rangelands

3

Graham Griffin

Introduction

Almost 300 million people of the world, across all continents (Fig. 3.1) are indigenous. The term is both spatially and temporally scale-dependent (all humans are indigenous to the earth). 'Indigenous' has come to mean the original inhabitants of an area that has been subsequently occupied by migrants (Seymour-Smith, 1986). Most colonization has been by migrants from western Europe over the last few centuries. Even this definition can be problematic, given the long history of repeated population movements and colonization world-wide. Frequently the term is reserved for populations which occupy an economically or politically marginal role compared with later arrivals, in what the 'indigenous' people regard as their own land. But far more people consider themselves indigenous than this. Peoples not affected by recent migration, but none the less having a very long history of occupation and association with land, number 1.5–2 billion. They are members of local communities practising traditional lifestyles, mostly in the rangelands of Africa, central Asia, the Americas, the Middle East and Australia (United Nations Environment Programme, 1992; United Nations Environment Programme, 1996).

Alienation and displacement of indigenous peoples over the past few centuries have created massive poverty, inequality and the degradation of traditional lifestyles. Many groups have been concerned about the persistence of their cultural and ethnic identity in the face of technological and population change. Debates about associated social

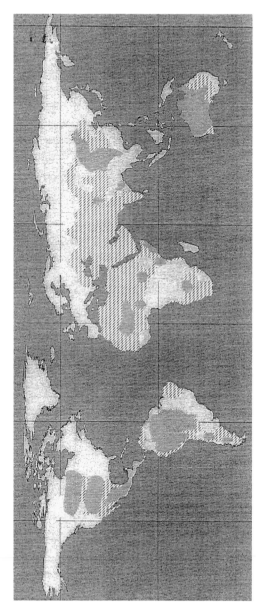

Fig. 3.1. Distribution of rangelands (hatched area) (adapted from the distribution of hyperarid, arid, semi-arid and dry sub-humid climate regions (United Nations Environment Programme, 1992)) and regional concentrations of indigenous peoples (shaded area) (as defined by Burger, 1990) who live in rangelands (adapted from Burger, 1990).

and environmental issues are often highly emotive and politically sensitive. This is partly due to contrasting world views, including competing perceptions of natural resources and appropriate resource use. Many indigenous people perceive themselves as intrinsic elements of the natural resources, able only to live within their means and not export elements of the environment to other places and people. This contrasts directly with many features of modern technological societies that wish to exploit resources for high levels of wealth generation by harvest and export. Indigenous people are increasingly asserting their right to own and control these resources and to return to less exploitative resource use (Schwartz, 1994; Peers, 1997). They are often seen as bastions of traditional lifestyles, being less technologically advanced, but, according to some, more environmentally friendly than land users from western societies.

It is frequently argued that non-industrial societies evolved sound subsistence strategies suited to their environment (see for example Hammett, 1992; Alcorn, 1993; Dwyer, 1994; Agrawal, 1995). However, *there is ample evidence of substantial damage to resources and land degradation caused by most human groups under most forms of land use* (Blaikie and Brookfield, 1987; Flannery, 1991; Kay, 1994; Kohen, 1995). Indeed, colonizing populations assumed that indigenous systems of resource management were primitive and unstructured, and hence were best replaced with what were perceived as modern and efficient resource uses (see for example Miller, 1999). Colonizing populations rarely regarded indigenous ways of using and valuing resources as significant. In addition to physical immigration, new technologies and different social and cultural ideologies have permeated many indigenous societies, particularly over the latter half of the last century, resulting in further alienation and displacement (Milton, 1999).

Most indigenous rangeland groups practised semi-subsistence hunting, gathering, grazing and/or seasonal agriculture. Compared with farming areas in higher rainfall regions, rangelands were variable and of low productivity, leading to flexible patterns of land use, rarely based on individual ownership or precisely circumscribed areas of occupation. Indigenous populations were usually of low density and mobile. However, these areas were among the most biologically diverse lands in the world (International Society of Ethnobotany, 1988) and included highly productive grazing land (FAO, 1980).

What changes were experienced by indigenous populations through colonization, displacement and incorporation into global economies? What problems arose when indigenous people competed for access to rangeland resources with people from different cultural or technological backgrounds? How are these problems recognized and articulated? Are there ways in which the inequalities and damage of the past and present might be redressed?

Colonization

In the vast majority of cases, *colonization curtailed indigenous people's land ownership and resource access, restricted alternative livelihoods and opportunities, and marginalized them politically.* A wide range of social, economic and environmental indicators show indigenous people to be among the world's poorest and most disadvantaged (FAO, 1993, 1996). While some colonizers tried to remove indigenous populations from the rangelands, most were prepared to coexist with them, under the immigrants' terms. Land was settled and used for commercial grazing, agriculture and mining. Indigenous people were often co-opted into the new enterprises as labour. Tenure and access conditions were altered to bring indigenous people into the new economies and limit their potential to resist further immigration (Brown and Jones, 1999). Even social and economic development programmes aimed at improving the well-being of indigenous populations often had profound impacts, as national governments and aid agencies perceived the traditional way of life as incompatible with modern values.

East African pastoralists have steadily lost land to farming and tourism developments over the last 30 years (Fratkin, 1999). In most areas, neighbouring peoples displaced local populations. Land was appropriated for commercial farming, funded by international aid agencies. Competition for land resources was exacerbated by population growth, drought, famine, commoditization, sedentarization, urban migration, political turmoil and civil war. In Namibia, centralization of land tenure control and state control of wildlife has alienated resources; as a result indigenous people lost their livelihood and wildlife populations declined (Brown and Jones, 1999).

The extent of colonization, displacement and disruption to indigenous populations over the past few centuries is illustrated in Table 3.1. The nature and impact of colonizations in different regions and continents has been vastly different. In some areas there have been attempts to completely remove indigenous populations from vast areas (Moore, 1989). In others the colonizers have attempted to coexist with indigenous people. Attempts to incorporate indigenous people into the broader economy and newer land use practices have failed (Fratkin, 1991, 1999; Milton, 1999). Where coexistence evolved, indigenous people remain marginalized and experience declining living conditions. To overcome these problems governments have privatized large areas of formerly communal rangelands. However, indigenous people have resisted attempts to restrict their activities to the newly tenured land parcels. Where local subsistence economies have been replaced, this has often had profound social and environmental effects, leading invariably to poorer people in socially and economically

Table 3.1. Colonization of lands and effects on the distribution and lives of indigenous herders, foragers and hunters.

Region	Immigration	Foragers and hunters	Herders	Population of indigenous peoples in rangelands (estimated from Burger, 1990)	References
Africa	Extensive mobility and colonization within regions and between ethnic groups. Recent colonization mainly from western Europe over last 300 years. Nation building by centralized States and civil wars continue to disrupt.	Extensively across most of Africa.	Across most of Africa, particularly in the north (Sahara and Sahel) and south. Some adaptation to new herding economies.	26 million	Schneider, 1979; Crummey and Stewart, 1981; Sandford, 1986; Galaty and Bonte, 1991; Barnard, 1992; Smith, 1992; Majok and Schwabe, 1996; Spencer, 1998; Lanyasunya *et al.*, 1999
Australia	Colonized from western Europe over last 200 years. Strong land rights development recovering land in remote regions.	Over the entire continent.	None. Some recent adoption of pastoralism in arid and tropic areas.	250,000	Peterson and Langton, 1983; Schrire and Gordon, 1985; Peterson and Long, 1986; Dingle, 1988; Lourandos, 1997
Central Asia	Massive internal population movements, displacement and immigration over millennia. Civil war, sedentarization and segregation continue.	Over most of Asia. Mostly confined now to far north, far south and south-east in mountainous regions.	Most of central Asia and northern and western areas of the former USSR. Extensive use in rangelands persists.	89 million	Smith, 1991; Bothe *et al.*, 1993; Minority Rights Group, 1994; Slezkine, 1994; Harris, 1996; Humphrey and Sneath, 1996, 1999; Tsundue, 1999

Continued

Table 3.1. *Continued*

Region	Immigration	Foragers and hunters	Herders	Population of indigenous peoples in rangelands (estimated from Burger, 1990)	References
Middle east and south Asia	Massive internal population movements, displacement and immigration over millennia. Civil war, sedentarization and segregation continue.	Few isolated groups persist, mainly in mountainous areas.	Continued extensive use of rangelands.	104 million	Behnke, 1980; Chatty, 1986; Maisels, 1990; Harris, 1996; Badjian and Baktiar, 1999
North America	Complete colonization from western Europe over last 500 years.	Across the entire continent.	Limited herding practised. Land rights evolving and recovering land, particularly in the far north.	1.5 million	Williams and Hunn, 1982; McNeil, 1983; Schwartz, 1986; Young *et al.*, 1991; Fixico, 1998; Marks, 1998; Ross, 1999
South America	Colonization of most regions over last 500 years. Indigenous populations now very small. Civil war, sedentarization and segregation continue.	Across the central and northern parts of the continent	Across the southern central and eastern regions.	3 million	Moore, 1989

marginalized communities. Not all indigenous peoples have been colonized in recent centuries and some have only experienced partial colonization. Some indigenous groups persist with their social and cultural systems still largely intact.

The continued persistence and sustainability of traditional land use systems does present many challenges. Government attempts over the last 30 years to nationalize Middle Eastern rangelands to displace herders' customary law (Rae *et al.*, 1999) have failed. Indigenous institutions for rangeland management persisted and ensured that use of the rangeland was sustainable. However, growing populations, land scarcity and demands for commercialization of production are placing enormous pressures on the land and on existing indigenous management systems. Diverse use of land by peasants and nomads in Iran appears to be effectively maintaining both the resources and people's livelihoods (Ansari, 1999; Badjian and Baktiar, 1999; Shahvali and Badjian, 1999). Tsundue (1999) also recorded the way in which traditional herding methods were adapted to the environment in Tibet but identified that international aid programmes targeting development have been detrimental to the persistence of indigenous lifestyles and practices. Indigenous systems and land use persist in Tibet in the face of enormous pressure for change by the Chinese government (Miller, 1999). While the competition for resource use is limited because of the harsh climate, the political pressure to sedentarize, commercialize and modernize land use practices is substantial. Likewise in Africa, Fratkin (1999) and Lanyasunya *et al.* (1999) demonstrated the viability of subsistence pastoralism and argue for economic and political strategies that support pastoral sustainability rather than displace it. Rae *et al.* (1999), pointing to the persistence and effectiveness of customary institutions, concluded that these represented substantial resources for policy-makers, if they were prepared to devolve management to local organizations.

Redressing the Balance

The emergence and spread of human and indigenous rights movements over recent decades has led to some redressing of inequalities (Kottak, 1999) *in some areas* (Australia, Canada and North America). Traditional knowledge about resources and environmental processes has contributed substantially to the understanding of the rangelands and their potential. Compensation for, and recognition of, the source of this knowledge is growing. There are also increasing attempts to restore land ownership or use rights to indigenous people in recently colonized countries. The few documented examples of access and use conflict resolution suggest that some success comes from mutually acknowledging and respecting each party's values and perceptions. This includes both resource identification and use (see for example Croll and Parkin, 1992; Price Cohen, 1998; Suksi, 1998; Havemann, 1999).

Despite the highest ideals and directives (United Nations High Commissioner for Human Rights, 1997)[1] many activists are concerned that *indigenous societies and subsistence practices are unlikely to persist in the face of economic globalization.* Concurrent with this is the sense that the loss of indigenous cultural (and, with it, biological) diversity threatens the persistence of humans, and, more generally, the diversity of life in the rangelands. While many indigenous peoples have resisted assimilation into the new societies, most have had no option but to substantially modify their lifestyles. *Changed population levels, social and environmental conditions, and lifestyle aspirations mean that it is not feasible for most indigenous peoples to return to their traditional lifestyles and land use practices.* Traditional rights, especially those maintained by warfare, invasion and social inequalities, are incompatible with most modern nation-state political ideologies. However, many indigenous cultures have shown remarkable resilience, and new lifestyles have emerged drawing on traditional skills. None the less, indigenous people require access to land and resources if they are to retain at least some elements of their culture, lifestyle and knowledge. National and international agreements and legislation to protect indigenous knowledge and resource use are developing (Table 3.1), but must tackle complex issues regarding knowledge and ownership (see for example Johannes, 1989; Brown, 1998; Fourmile, 1998).

Future

New strategies for understanding different perspectives, values and involvement in rangeland resource use are emerging in some areas. Participatory rural appraisal and rapid ecological assessment programmes are reversing the effects of over a century of colonialism in Namibia (Brown and Jones, 1999). In parts of Africa, communal-area conservation agreements provide opportunities to manage resources such as grazing by domestic stock, wood products and water (Matzke and Nabane, 1996; Pilotlight, 1998). Significant areas of land have been returned to Aboriginal ownership in the arid rangelands of Australia. On Aboriginal-owned lands, Indigenous Protected Area agreements

---

[1] Article 14: 'The rights of ownership and possession of the peoples concerned over the lands which they traditionally occupy shall be recognized,' and, Article 15: 'to participate in the use, management and conservation of these resources'.

between government and indigenous people are tackling conservation issues (Noble and Ward, 1999). Participatory planning strategies have set land use strategies based on Aboriginal priorities in Australian rangelands (Davies *et al.*, 1999). Service delivery programmes developed for and by indigenous people benefit cultural aspects of American Indian communities on their own lands (Tippenconic Fox and Stauss, 1999). Recognizing and incorporating traditional ecological knowledge in resource management as well as providing a mechanism for indigenous people to benefit from the use of their knowledge are positive new developments in some countries. Whilst such approaches are relatively recent and localized, they demonstrate the possibility of national programmes supporting rather than undermining indigenous people's use of rangeland resources. While traditional rangeland science may not have a lot to offer in this process it will be critical for rangeland scientists to include the recognition of differing human values and perceptions into their understanding of resources and resource uses. *Rangeland scientists should, like members of the broader society, recognize the variety of valid perceptions of and involvements in rangelands and their resources and not perceive western technological land use and commercial pastoralism as the only or best option for the persistence and sustainability of the rangelands and the people who live in them.*

Acknowledgements

I am grateful to Sarah Dunlop for critically reviewing an early draft of this chapter and suggesting many improvements.

References

Alcorn, J.B. (1993) Indigenous peoples and conservation. *Conservation Biology 7*, 424–426.

Agrawal, A. (1995) Dismantling the divide between indigenous and scientific knowledge. *Development and Change*, 26, 413–439.

Ansari, N. (1999) Traditional rangeland utilization in the western region of Iran. In: Eldridge, D. and Freudenberger, D. (eds) *People of the Rangelands. Building the Future. Proceedings of the VI International Rangeland Congress*. VI International Rangeland Congress Inc., Townsville, Australia, p. 406.

Badjian, G.R. and Baktiar, A. (1999) The rule of rural beneficiaries in indigenous rangeland management. In: Eldridge, D. and Freudenberger, D. (eds) *People of the Rangelands. Building the Future. Proceedings of the VI International Rangeland Congress*. VI International Rangeland Congress Inc., Townsville, Australia, p. 409.

Barnard, A.J. (1992) *Hunters and Herders in Southern Africa: a Comparative Ethnography of the Khoisan Peoples*. Cambridge University Press, Cambridge, UK.

Behnke, R.H. (1980) *The Herders of Cyrenaica: Ecology, Economy, and Kinship Among the Bedouin of Eastern Libya*. University of Illinois Press, Urbana, Illinois.

Blaikie, P. and Brookfield, H. (1987) *Land Degradation and Society*. Methuen, London.

Bothe, M., Kurzidem, T. and Schmidt, C. (1993) *Amazonia and Siberia: Legal Aspects of the Preservation of the Environment and Development in the Last Open Spaces*. Graham & Trotman, London.

Brown, C.J. and Jones, B.T.B. (1999) Common-property rangelands management in Namibia: the 'conservancy' model in communal areas. In: Eldridge, D. and Freudenberger, D. (eds) *People of the Rangelands. Building the Future. Proceedings of the VI International Rangeland Congress*. VI International Rangeland Congress Inc., Townsville, Australia, pp. 411–413.

Brown, M.F. (1998) Can culture be copyrighted? *Current Anthropology* 39, 193–222.

Burger, J. (1990) *The Gaia Atlas of First Peoples*. Penguin Books, Ringwood, Australia.

Chatty, D. (1986) *From Camel to Truck: the Bedouin of the Modern World*. Vantage, London.

Croll, E. and Parkin, D. (1992) Cultural understandings of the environment. In: Croll, E. and Parkin, D. (eds) *Bush Base: Forest Farm. Culture, Environment and Development*. Routledge, London, pp. 11–36.

Crummey, D. and Stewart, C.C. (1981) *Modes of Production in Africa: a Precolonial Era*. Sage Publications, Beverly Hills, USA.

Davies, J., Higginbottom, K., Noack, D., Ross, H. and Young, E. (1999) *Sustaining Eden: Indigenous Community Wildlife Management in Australia*. International Institute for Environment and Development, London.

Dingle, A.E. (1988) *Aboriginal Economy: Patterns of Experience*. McPhee Gribble Penguin, Melbourne, Australia.

Dwyer, P.D. (1994) Modern conservation and indigenous peoples in search of wisdom. *Pacific Conservation Biology* 1, 91–97.

FAO (1980) *Population and Agriculture in the Developing Countries. Economic and Social Development Papers*. Food and Agricultural Organization, Paris.

FAO (1993) *Intercountry Comparisons of Agricultural Output and Productivity. Economic and Social Development Papers*. Food and Agricultural Organization, Paris.

FAO (1996) *Economic Development and Environmental Policy. Economic and Social Development Papers*. Food and Agricultural Organization, Paris.

Fixico, D.L. (1998) *The Invasion of Indian Country in the Twentieth Century; American Capitalism and Tribal Natural Resources*. University of Colorado Press, Niwot, Colorado.

Flannery, T. (1991) The mystery of the Meganesian meat-eaters. *Australian Natural History* 23, 722–729.

Fourmile, H. (1998) Using prior consent procedures under the Convention on Biological Diversity to protect indigenous traditional ecological knowledge and natural resource rights. *Indigenous Law Bulletin* 4, 14–15.

Fratkin, E. (1991) Pastoralism: governance and development issues. *Annual Review of Anthropology* 26, 235–261.

Fratkin, E. (1999) The survival of subsistence pastoralism in east Africa. In: Eldridge, D. and Freudenberger, D. (eds) *People of the Rangelands. Building the Future. Proceedings of the VI International Rangeland Congress.* VI International Rangeland Congress Inc., Townsville, Australia, pp. 392–396.

Galaty, J.G. and Bonte, P. (1991) *Herders, Warriors, and Traders: Pastoralism in Africa.* Westview Press, Boulder, Colorado.

Hammett, J.E. (1992) The shapes of adaptation – historical ecology of anthropogenic landscapes in the southeastern United-States. *Landscape Ecology* 7, 121–135.

Harris, D.R. (1996) *The Origins and Spread of Agriculture and Pastoralism in Eurasia.* UCL Press, London.

Havemann, P. (1999) *Indigenous Peoples' Rights in Australia, Canada, and New Zealand.* Oxford University Press, Oxford, UK.

Humphrey, C. and Sneath, D. (1996) *Culture and Environment in Inner Asia.* White Horse, Cambridge, UK.

Humphrey, C. and Sneath, D. (1999) *The End of Nomadism? Society, State, and the Environment in Inner Asia.* Duke University Press, Durham, North Carolina.

International Society of Ethnobotany (1988) The Declaration of Belem. http://users.ox.ac.uk/~wgtrr/belem.htm.

Johannes, R.E. (1989) *Traditional Ecological Knowledge: a Collection of Essays.* IUCN Publication Services, Gland, Switzerland.

Kay, C.E. (1994) Aboriginal overkill – the role of native Americans in structuring western ecosystems. *Human Nature – an Interdisciplinary Biosocial Perspective* 5, 359–398.

Kohen, J. (1995) *Aboriginal Environmental Impacts.* University of New South Wales Press, Sydney, Australia.

Kottak, C.R. (1999) The new ecological anthropology. *American Anthropologist* 101, 23–35.

Lanyasunya, T.P., Ole Sinkeet, S.N., Mukisira, E.A. and Siamba, D.N. (1999) Range–livestock interface, perturbations and implication in Kenya. In: Eldridge, D. and Freudenberger, D. (eds) *People of the Rangelands. Building the Future. Proceedings of the VI International Rangeland Congress.* VI International Rangeland Congress Inc., Townsville, Australia, pp. 407–409.

Lourandos, H. (1997) *Continent of Hunter-gatherers: New Perspectives in Australian Prehistory.* Cambridge University Press, Cambridge, UK.

Maisels, C.K. (1990) *The Emergence of Civilisation: From Hunting and Gathering to Agriculture, Cities, and the State in the Near East.* Routledge, London.

Majok, A.A. and Schwabe, C.W. (1996) *Development Among Africa's Migratory Pastoralists.* Bergin & Garvey, Westport, Connecticut.

Marks, P.M. (1998) *In a Barren Land: American Indian Dispossession and Survival.* William Morrow, New York.

Matzke, G.E. and Nabane, N. (1996) Outcomes of a community controlled wildlife utilization programme in a Zambezi valley community. *Human Ecology* 24, 65–85.

McNeil, K. (1983) *Indian Hunting, Trapping and Fishing Rights in the Prairie Provinces of Canada.* University of Saskatchewan, Saskatoon, Canada.

Miller, D. (1999) Herders of forty centuries: nomads of Tibetan rangelands in western China. In: Eldridge, D. and Freudenberger, D. (eds) *People of the Rangelands. Building the Future. Proceedings of the VI International Rangeland Congress.* VI International Rangeland Congress Inc., Townsville, Australia, pp. 402–403.

Milton, K. (1999) Colonization and local populations: the rangeland experience. In: Eldridge, D. and Freudenberger, D. (eds) *People of the Rangelands. Building the Future. Proceedings of the VI International Rangeland Congress.* VI International Rangeland Congress Inc., Townsville, Australia, pp. 387–391.

Minority Rights Group (1994) *Polar Peoples: Self Determination and Development.* Minority Rights Publications, London.

Moore, K.M. (1989) Hunting and the origins of herding in Peru. PhD Thesis, University of Michigan.

Noble, K. and Ward, I. (1999) Ngaanyatjarra – traditional people, contemporary conservation. In: Eldridge, D. and Freudenberger, D. (eds) *People of the Rangelands. Building the Future. Proceedings of the VI International Rangeland Congress.* VI International Rangeland Congress Inc., Townsville, Australia, pp. 405–406.

Peers, D.M. (1997) *Warfare and Empires. Contact and Conflict Between European and Non-European Military and Maritime Forces and Cultures.* Expanding World, Vol. 24. Ashgate/Variorum, Brookfield, Vermont.

Peterson, N. and Langton, M. (1983) *Aborigines, Land and Land Rights.* Australian Institute of Aboriginal Studies, Canberra, Australia.

Peterson, N. and Long, J. (1986) *Australian Territorial Organization: a Band Perspective.* University of Sydney, Sydney, Australia.

Pilotlight (1998) Pilotlight. http://www.whoseland.com/about.html

Price Cohen, C. (1998) *The Human Rights of Indigenous Peoples.* Transnational Publishers, New York.

Rae, J., Arab, G. and Nordblom, T. (1999) Customary regulation of steppe pastures in Syria. In: Eldridge, D. and Freudenberger, D. (eds) *People of the Rangelands. Building the Future. Proceedings of the VI International Rangeland Congress.* VI International Rangeland Congress Inc., Townsville, Australia, pp. 396–401.

Ross, H. (1999) New ethos, new solutions: lessons from Washington's co-operative environmental management agreements. *Australian Indigenous Law Reporter* 4, 1–28.

Sandford, S. (1986) Traditional African range management practices. In: *Rangelands: a Resource Under Siege. Proceedings of the 2nd International Rangelands Congress.* Australian Academy of Science, Canberra, Australia, pp. 474–478.

Schneider, H.K. (1979) *Livestock and Equality in East Africa: the Economic Basis for Social Structure.* Indiana University Press, Bloomington, Indiana.

Schrire, C. and Gordon, R. (1985) *The Future of Former Foragers in Australia and Southern Africa.* Cultural Survival Inc., Cambridge, Massachusetts.

Schwartz, B. (1986) *First Principles, Second Thoughts: Aboriginal Peoples, Constitutional Reform and Canadian Statecraft.* Institute for Research on Public Policy, Montreal, Canada.

Schwartz, S.B. (1994) *Implicit Understandings Observing, Reporting, and Reflecting on the Encounters between Europeans and Other Peoples in the Early Modern Era.* Studies in Comparative Early Modern History. Cambridge University Press, New York.

Seymour-Smith, C. (1986) *Macmillan Dictionary of Anthropology.* Macmillan Press, London.

Shahvali, M. and Badjian, G.R. (1999) Reconstruction of indigenous rangeland management in the south-west of I.R. Iran. In: Eldridge, D. and Freudenberger, D. (eds) *People of the Rangelands. Building the Future. Proceedings of the VI*

International Rangeland Congress. VI International Rangeland Congress Inc., Townsville, Australia, p. 410.

Slezkine, Y. (1994) *Arctic Mirrors: Russia and the Small Peoples of the North*. Cornell University, New York.

Smith, A.B. (1992) *Pastoralism in Africa: Origins and Development Ecology*. Christopher Hurst, London.

Smith, E.A. (1991) *Inujjuamit Foraging Strategies: Evolutionary Ecology of an Arctic Hunting Economy*. A. de Gruyter, New York.

Spencer, P. (1998) *The Pastoral Continuum: the Marginalisation of Tradition in East Africa*. Oxford University Press, Oxford, UK.

Suksi, M. (1998) *Autonomy Applications and Implications*. B. Kluwer Law International, Boston, Massachusetts.

Tippenconic Fox, M. and Stauss, J.H. (1999) Cooperative extensions of indigenous people: a unique experience? In: Eldridge, D. and Freudenberger, D. (eds) *People of the Rangelands. Building the Future. Proceedings of the VI International Rangeland Congress*. VI International Rangeland Congress Inc., Townsville, Australia, p. 404.

Tsundue, K. (1999) Pastoral-nomadism in Tibet: between tradition and modernization. In: Eldridge, D. and Freudenberger, D. (eds) *People of the Rangelands. Building the Future. Proceedings of the VI International Rangeland Congress*. VI International Rangeland Congress Inc., Townsville, Australia, pp. 413–415.

United Nations Environment Programme (1992) Convention on Biological Diversity. No. 92–7807.

United Nations Environment Programme (1996) Convention to combat desertification in countries experiencing serious drought and/or desertification, particularly in Africa.

United Nations High Commissioner for Human Rights (1997) 'Convention (No. 169) concerning indigenous and tribal peoples in independent countries.' Web page. Available at http://www.unhchr.ch/html/menu3/b/62.htm.

Williams, N. and Hunn, E. (1982) *Resource Managers: North American and Australian Hunter-gatherers*. AAAS, Washington, DC.

Young, E., Ross, H., Johnson, J. and Kesteven, J. (1991) *Caring for Country: Aborigines and Land Management*. Australian National Parks and Wildlife Service, Canberra, Australia.

Rangelands: People, Perceptions and Perspectives

4

Denzil Mills, Roger Blench, Bertha Gillam, Mandy Martin, Guy Fitzhardinge, Jocelyn Davies, Simon Campbell and Libby Woodhams

A very important question is, 'What do people of rangelands want landscapes they live and work in to look like?' Rangeland landscapes have always been 'managed' in some form, and have mostly slowly changed through time. Are the people of the rangelands going to allow this change to continue, or are they going to manage for some preferred 'state'? How do they deal with the notion of pristine, and if it is accepted that there can never be a 'pristine state', then where should the line be drawn?

Sustainability is discussed, but what does it look like, given that it is a process rather than an outcome? We need to consider the question of what state is to be sustained. Given that social systems and ecosystems are seen as incompatible by some, and that 'most [modern environmentalists] equate productive work in nature as destruction' (White, 1996), there is obviously going to have to be compromise. Scientists and people of rangelands appear never to ask the questions, 'What do we want the rangelands to look like?'; 'Who do we want to be there?'; and 'What do we want them to be doing?' One attempt has been made to define these issues in Australia (National Rangeland Management Strategy, 1996) by recording in some detail the community expectations for use of rangelands. However, in spite of the progress made by this report, it is still ignored as a foundation on which to build future action.

What is this need to 'dialogue with the landscape'; an urge which, we believe, is shared by most rangeland people whether they be indigenous, scientist, grazier, institution or even urban people?

©CAB International 2002. Global Rangelands: Progress and Prospects
(eds A.C. Grice and K.C. Hodgkinson)

Background

In managing rangelands, regardless of the scientific information available to them, people draw, either consciously or unconsciously, on knowledge that derives from their memes and relationships with particular places, ecosystems and landscapes. Memes are ideas, habits, skills, behaviours, inventions, songs and stories copied from other humans, especially parents and ancestors (Dawkins, 1976; Blackmore, 2000). Rangeland cultures are made up of memes and are inseparable from local knowledge systems, having co-evolved with them.

Rangeland landscapes are the product of local knowledge and practice. Yet the significant role of local knowledge in rangeland decision-making is rarely recognized (but see Heywood *et al.*, 2000) because such knowledge often fails to fit the conventional scientific paradigms of rangeland management, being contextual, value-laden, holistic and not readily amenable to reductionist analysis. Hence local knowledge is not readily understood if transported to other contexts.

Local knowledge is the basis for creative and spiritual expression of attachment to place, through a variety of forms such as painting, song, dance, literature and oral stories largely passed on by imitation. It is also the basis on which people learn about, manage and monitor the state of their general environment and specific landscapes. Recently, increasing attention to the traditional ecological knowledge of rangeland peoples has highlighted differences between local and scientifically based knowledge and indicated the potential contributions of local knowledge to strategies for sustainable development in the rangelands.

All rangelands occupy a unique place in the ethos and context of nations that have rangelands within their borders. For example, Australians see themselves as a nation that has been built on myths and legends forged mostly in the arid and semi-arid heartlands: the rangelands. Although most Australian people live outside the rangelands, there is a connection, often mythical, with the lives of 'stockmen'. Also the urban people outside the rangelands believe the special lands are being degraded and this sets up tensions that cross many scales, communities and institutions.

In the forefront of the battle to balance community expectations with ecological necessity are such issues as threats to the health and sustainability of the ecosystem through inappropriate use, low levels of economic and social capital and the diminishing rangeland communities. The environment in which these decisions take place is one of sentimentality devoid of experience; the tendency to look at landscape and community as separate issues, and the persistent application of inappropriate understandings. A holistic approach to rangeland management that includes the requirements of both social systems and natural ecosystems is urgently required (Fitzhardinge, 1999).

People

The most immediate and pressing problems for rangeland management are 'people problems'.

Rangeland communities throughout the world suffer enormously from isolation. Isolation in this context includes geographical, temporal, political and perceptual isolation. This 'rural/urban distinction under-lies many of the power relations that shape the experiences of people in nearly every culture' (Ching and Creed, 1997). Nowhere is this becoming more of a reality than in the rangeland areas of Australia. The depressed socio-economic environment of most regions contrasts markedly with the booming national economy and the ethos of global-ization, youth, multiculturalism and conspicuous consumption that characterizes cities. Changes in demography and pastoral industry economics have meant a loss of political and economic power. All this has led to a loss of what Pretty calls 'social capital' (Pretty, 1997). The reduction of services, of schools and educational facilities, of communication facilities, transport and other facilities has led to fur-ther isolation. The concept of social capital is especially important in the context of recognizing the need for and dealing with change. A community's ability to recognize the need for change and to manage the process would appear to be strongly related to the strength and matu-rity of that community. This is akin to Pretty's social capital.

Perceptions

Different understandings about these problems may result from the different values, objectives and methods in different knowledge systems.

Local knowledge derives from people's direct experience of the distinctive social and physical character of particular places and is underpinned by people's attachments to those places. It can be defined as comprising 'the categories, meanings and cultural practices that "local" people use to make sense of their world' (Murdoch and Clark, 1994). Local knowledge systems are thus ways of seeing the landscape, which both reflect and shape social values and local people's uses and management of land. They are 'learned ways of knowing and looking at the world. They have evolved from years of experience and trial and error problem solving by groups of people working to meet the challenges they face in their local environments, drawing upon the resources at hand' (McClure, 1989, in De Walt, 1994).

Local knowledge also embodies claims to power over land and resources, especially in the face of counter claims from 'outsiders'. Hence, when noted Aboriginal artist from the Walmajarri Tribe, Jimmy Pike, learned that the designation of his country in north-western Australia as Vacant Crown Land meant that it belonged to the Queen, he is reported to have declared: 'The Queen never bin fuggin walk around here! Bring her here and I'll ask her: All right, show me all the waterholes!' (Lowe, 1997).

However, although indigenous peoples have local knowledge systems of long standing, it is a mistake to equate local knowledge only with traditional or indigenous knowledge (De Walt, 1994). All people have local knowledge, though clearly not all local knowledge is the same. Just as each local 'community' comprises a mix of often diverse and disparate social groups and cultures, so local knowledge systems can present varied and contested ways of knowing and of managing the same local environments, all of which may be different from those of 'outsiders'.

Such is the situation in the Australian rangelands, where local knowledge systems of Aboriginal people and graziers contest with each other for power as well as interfacing variously with the 'outsider' perspectives of conservationists, government officials and scientists. This contestation limits the opportunities for sustainable development to build from the local knowledge systems of both Aboriginal people and graziers.

What Aboriginal people and graziers share are inextricable linkages of people and landscape in their respective cultures. Each group's local knowledge has co-evolved with their uses of landscape.

Culture in the rangelands embraces both first and second settler histories and readings of landscape.

However, European settlers 'didn't know the landscape'. There is not a long history of intergenerational experience, and when Europeans came to Australia they devalued the knowledge of those who did know the landscape. The first settlers came with expectations derived in another part of the world, and a language, culture and 'knowledge' that made little sense in their newfound home. Perceptions of landscape in many cases were dependent on what people expected to find (for example, Martin, 1993; Seddon, 1997). Even the term 'landscape' itself is a cultural invention. As Seddon says, it was the 'explanation of the unfamiliar by the familiar' (Seddon, 1997).

This has had ramifications for the slow evolution of scientific understandings of the Australian ecosystem and its sustainable management. It has also influenced the development of art. Using paintings, stories and other art forms as readings of landscape promotes social cohesion. It reinforces common values and allows points of conflict to become apparent in non-threatening ways.

Art, or creative activity in all its manifestations, is a prime mode of communication in local knowledge systems. For settler cultures in Australia, artistic expression reflects the development of under-standings and attachments to place.

As George Seddon comments, drawing from Brian Elliott's succinct reasoning, in his essay 'Sense of Place' behind the need for art:

> The first need in a new country or colony must obviously be in one way or another to comprehend the physical environment. In poetry we find this need reflected, in colonial times, in an obsessive preoccupation with landscape and description. At first the urge is merely topographical, to answer the question, what does the place look like? The next is detailed and ecological: how does life arrange itself there? What plants, what animals, what activity? The next may be moral: how does such a place influence people? And how, in turn, do the people make their mark upon the place? How have they developed it? Next come subtler enquiries: what spiritual and emotional qualities do such a people develop in such an environment? In what way do the forces of nature impinge upon the imagination? How do aesthetic evaluations grow? How may poetry come to life in such a place as Australia?
>
> The most important parallel between language and the physical environment is that both function as media, whether explicable or not, for complex symbolic expression. Although learning in the object world occurs both before and during language acquisition, cognitive processes have largely been analyzed through linguistic models that present language as the primary determinant of perception, and suggest that cultural identity requires mutual agreement on categories and concepts. Such linguistic models fail to adequately explain how the physical, visual and non-verbal symbolic universe contributes to cognitive interaction. Recently, linguistics, psychology and anthropology have begun to correct this imbalance, acknowledging fully the object world's role as 'vehicles of meaning' (Miller, 1987). As Piaget (1951) has pointed out, language is incorporated at a relatively late developmental stage in constructing a symbolic universe. In addition, language is often limited by its explicit and linear form, requiring consecutive interpretation, while visual or physical symbolism is more lateral in its nature, permitting simultaneous absorption of a complex of meanings. Thus the physical and visual worlds, rather than language, are the primary media in which every object or image carries meanings, associations and values, which may be expressed through language, ritual, art and action.

(Seddon, 1997)

This discussion of the role of the non-verbal helps account for the use of objects and topography, as in song lines or Polynesian shell-star maps, to define relationships and meaning, not without words, but using words as the cement in the building blocks of knowledge and values attributed to the landscape.

For the attainment of sustainable systems, both the ecosystem and the socio-economic systems must be healthy.

Agricultural systems as introduced to Australia from England and Europe were a product of a long co-evolution between the social system and the ecosystem; an evolution that took place in another part of the world. For those who doubt the importance of this fact, it is worth contemplating how Australia would look had the land been colonized by Arabs, for example. No doubt we would have a different balance of imported animals, a different land tenure system, and perhaps even a different financial system, remembering that under Islam the payment of interest is haram (anathema) (Buchan, 1997)!

How has this cultural preconditioning and lack of co-evolution with the ecosystem manifested itself in the rangelands? The impact of previously developed cultural mores has penetrated almost every area. It affects the land tenure system, with a current manifestation of that for Australian landholders being the challenges posed by Aboriginal land rights. It affects how properties are run and managed (with introduced sheep and cattle, 'drought' assistance). As a flow on from this, it affects how we value land ('What is it good for?') and other natural resources, such as water. However, probably the most damaging outcome in terms of the ecological and social sustainability of the rangelands has come from an unlikely source: the financial system.

In simple terms, the financial system enables the consumers of goods and services to reward the providers of these in a way that is a measure of their usefulness (value). Thus, graziers with flocks of sheep for wool production were rewarded for their labours by the price of wool. In terms of the wool production system, everything had a value in terms of the value of the final product. Things such as land, water, trees, biodiversity and the like could now be valued (in terms of their contribution to production). This then sent a clear message in a very practical sense to graziers about the 'ecosystem value' the community placed on natural resources. Levels of resource use were justifiable in terms of the financial rewards provided. It is only recently that there has been a general recognition of the problems this approach presents.

The reasons are clear. Monetary value is a very poor measure of real 'value' in all situations outside the financial system. It is hard to value biodiversity, for example. With the growing awareness of the ecological vulnerability of the Australian landscape, there is a growing raft of concerns and sentiments held by the larger community that are not measurable in a strictly monetary sense. This is the real hub of much of the problem of rangelands management. Individual graziers are being rewarded for doing one thing, but being asked to do something else. It will not be until there is a closer alignment of rewards with appropriate or desirable behaviour (as decided by the wider community) that substantial and sustainable changes in practice will be made.

> *Respect for all forms of local knowledge is important for social cohesion, regardless of the perceived utility of that knowledge in promoting production or ecological goals such as biodiversity conservation.*

Epistemological explorations of local knowledge – that is, examinations of its nature, methods and limitations – often compare it to science, drawing out similarities and differences. Science is presented as an objective way of knowing, free of contamination from social factors such as values (Clark and Murdoch, 1997). Its methods aim to remove the influence of the emotions, feelings, perceptions and intuition – the factors that determine the way that meaning is constructed out of experience in local knowledge systems (Kersten and Ison, 1994). Science is claimed to be powerful because its understandings are universally applicable whereas local knowledge can only achieve understanding in the particular contexts in which it was developed (Kloppenburg, 1991; De Walt, 1994; Murdoch and Clark, 1994).

However, attempts to rigidly distinguish scientific and local knowledge are flawed. Agrawal (1995) argues that both scientific and local knowledge systems are too heterogeneous to maintain a dichotomy and, further, that long-standing exchange and transformation of knowledge between cultures has now created pervasive interconnections. Recognizing that local knowledge is context dependent, Agrawal points out that scientific knowledge also has a context and that failure to appreciate this is a reason why technological solutions have so often failed when applied in rural development.

Science's claim that it is unique in being a system of value-free knowledge can be readily criticized, for example, because science has privileged value systems that see people as rightfully dominant over nature and has excluded values of harmony and cooperation. As feminists have pointed out, it reinforces patriarchal values. Indeed, what science studies is a product of tradition, fashion and other social, political and economic factors, and of practical logistics. These things determine what proposals will attract funding, what methods will be acceptable to peers and what outcomes scientists will seek. There is little that is objective or value-free about them (Kloppenburg, 1991; Murdoch and Clark, 1994; Turnbull, 1997).

The capacity of science to generate knowledge that is transportable, allowing it to be applied in many different contexts, is actually a function of how scientific knowledge is validated and transmitted, not of scientific modes of investigation. Science is, in fact, produced locally. It shares this localism with all other knowledge systems. Science starts with the detailed study of local phenomena – it then faces the problem of how to make itself universal (Clark and Murdoch, 1997).

All information or insight must be communicated to and accepted by other people before it becomes part of a system of knowledge. In local knowledge systems, information is disseminated using stories, art, ceremony, ritual, community associations, cooperative work practices and social gatherings. These build and reinforce the trust and authority needed for any information to be accepted as valid knowledge (Turnbull, 1997; Weeks and Packard, 1997). Science works in different ways. It creates models (temporal and spatial measurement, maps, algorithms, taxonomic hierarchies and dichotomous keys) to structure the messy nature of information in ways in which it can be more consistently described. In this sense it is the same as producing an art work – you put 'form' around a mass of perceptions ('seeing') and structure them into a picture that can be 'read'. In doing so, you generally 'pull out' the significant features – simplify to get the whole picture. Science uses methods such as disciplinary societies, peer review, published journal articles and reproducible experiments to validate knowledge and transmit it from one local setting to another (Clark and Murdoch, 1997; Turnbull, 1997). It is this well-developed impersonal capacity for organizing, validating and transporting knowledge through space and time that allows science to claim universality.

The legacy of the ideology that science is somehow better and more 'fit for survival' than local knowledge systems remains apparent in the 'deficit models' applied to rural development and extension activities. Deficit models conceive that people behave in ways that are undesirable (to the outside observer) because, being dependent on local knowledge, they lack the proper information (Weeks and Packard, 1997). Thus in environmental management, extension activity has frequently assumed that farmers mismanage soil and water as they lack the information on how to manage natural resources properly: 'What rural people know is assumed to be "primitive" and "unscientific", and so formal research and extension must "transform" what they know in order to "develop" them' (Pretty and Shah, 1997). Only recently has mainstream science and public policy in Australia recognized value in local knowledge systems and sought to work with them. One example of this is found in the Landcare Movement.

Landcare can be characterized as a social movement – since it transforms knowledge as people take action – 'cooperating, sharing, combining knowledge – to overcome the limits on the knowledge that they individually possess' (Wainright, 1994 in Hassanein and Kloppenburg, 1995). A significant proportion of graziers, perhaps 50%, are now involved in a Landcare or similar group.[1] Involvement seems to be positively correlated with education levels (Holmes and Day, 1995), suggesting the possibility that the more exposure that graziers have had to scientific concepts, the more likely they are to become involved in these participatory learning and action processes.

While innovative and creative, the participatory processes that are now active in Australian rangelands remain constrained by the conservatism of tradition. Grazier landcarers aim to establish more viable industries and higher quality of production. They are accommodating some conservation objectives but are unable or uninterested in redefining their role to be land stewards first and production managers second, as Holmes (1996) and Stafford Smith (1994) suggest is needed for sustainable management, at least in the many rangeland regions which do not have high future potential for grazing businesses. Conservatism also entrenches lack of effective communication between Aboriginal people and non-Aboriginal graziers. Differences in rangeland production systems, social and professional networks and communication styles isolate the participatory learning and action processes that the two groups are engaged in. Economic stress and a lack of creative thinking applied to rangeland tenures (Holmes, 1996) contribute to this conservative environment for planning. Even in regions where Aboriginal people and graziers are both accessing the same suite of government programmes to fund their participatory learning and action activities, there is little or no interface between what each group is doing and their ways of knowing.

Significant values and ways of seeing now held in common by non-Aboriginal graziers (Holmes, 1986; Holmes and Day, 1995) and Aboriginal people include their shared identity with rangeland landscapes, their sense of stewardship, and the way that their own personal experiences structure their readings of the landscape. Such commonalities suggest there is some basis for graziers and Aborigines to apply their local knowledge together in sustainable development of the rangelands. Social sustainability in the rangelands depends on this since, outside the towns, Aboriginal communities and pastoral

---

[1] Limited data are available on Landcare participation in the rangelands but 55% of the South Australian pastoralists who responded to Holmes and Day's (1995) survey are members of a Landcare group or similar organization. Critical comparison of Landcare in the rangelands and elsewhere would be valuable and is likely to reveal differences in participation and in the nature of Landcare activities which reflect biophysical and socio-cultural differences between rangelands and other rural regions. For example, as Baker (1997, p. 65) points out, it is probably no coincidence that many successful Landcare groups are found in small river valleys. In such places geography imparts a strong sense of community. In contrast, rangeland landscapes and land ownership patterns tend to disperse people. The high incidence of statutory organizations (e.g. Soil Conservation Boards in South Australia; Land Conservation District committees in Western Australia) amongst rangeland Landcare and similar groups suggests that government has needed to catalyse participatory learning and action in rangelands to a far greater extent than in other regions.

families are typically the only rangeland residents. However, recent power struggles over native title rights have imposed further barriers to doing so.

Science is most effective in promoting change where it builds on adaptive capacity.

There remain critically useful areas for science to be applied to link local knowledge into larger frameworks, and of course, vice versa. Sufficient research into learning theory and application in various group techniques makes it clear that local or indigenous knowledge is often accurate and should be valued along with the external data – but it cannot be admitted if it is patently illogical or physically incorrect. This includes learning that is described as 'heuristic', which is often held up as an important element of the learning obtained through group, adult or experiential learning systems. A useful definition of heuristics is 'specific mental strategies to solve specific problems' or a heuristic as 'a simple and approximate rule which solves a certain class of problems' (Piatelli-Palmarini, 1994).

It is important to note that it is not a question of the nature or the accuracy of the base information available to the person making a decision that produces the errors, *but the manner in which the person has learned to process or decide about the available information.* The key is providing the skills for people to 'think through' their opinions and decisions – to identify the roots of preferences and prejudices to produce a 'map' of the principles that are being invoked. For example, in clearing timber beside the road in the front paddock, family members were distressed not because it destroyed the koala's trees but because it destroyed the landscape of the father's story concerning his first arrival at the property.

The answer is not more information but 'friendly' ways to construct new heuristics that are more accurate. This suggests that an ideal and very important function for modelling/science is the development and presentation of heuristics (or 'rules of thumb') that are logical and integrated with a wider system or knowledge framework.

The Future – Integrating Knowledge

Key challenges for the future are to maintain the adaptive, resilient capacity of local knowledge systems, to integrate science with local knowledge in ways that promote sustainability and to encourage a system that rewards sustainable production, rather than production.

Strategies that might be adopted to address these important challenges include integration of local knowledge within educational systems and paying particular attention to language issues that inhibit communication. Extension strategies that inform and facilitate rather than rely on prescription should be encouraged, and the context of the scientific knowledge that is being used should be understood and available. The Australia National Rangeland Management Strategy (1996) should be adopted as a foundation on which to move forward in sustainable rangeland management in Australia. Promotion of awareness of outcomes from integration and the use of cultural media such as art to reflect on and challenge understandings are also strategies worth considering.

Finally, it is important to provide non-threatening forums to people of all backgrounds and ages to channel their knowledge into meaningful discussion of environmental issues.

References

Agrawal, A. (1995) Dismantling the divide between indigenous and scientific knowledge. *Development and Change* 26, 413–439.

Baker, R. (1997) Landcare: policy, practice and partnerships. *Australian Geographical Studies* 35, 61–73.

Blackmore, S. (2000) *The Meme Machine*. Oxford University Press, Oxford.

Buchan, J. (1997) *Frozen Desire. An Inquiry into the Meaning of Money*. Picador, London, UK.

Ching, B. and Creed, G.W. (eds) (1997) *Knowing your Place. Rural Identity and Cultural Hierarchy*. Routledge, New York.

Clark, J. and Murdoch, J. (1997) Local knowledge and the precarious extension of scientific networks: a reflection on three case studies. *Sociologia Ruralis* 37, 38–60.

Dawkins, R. (1976) *The Selfish Gene*. Oxford University Press, Oxford (revised edition with additional material, 1989).

De Walt, B.R. (1994) Using indigenous knowledge to improve agriculture and natural resource management. *Human Organization* 53, 123–131.

Fitzhardinge, G. (1999) People and landscape: growing together or growing apart? In: Eldridge, D. and Freudenberger, D. (eds) *People of the Rangelands. Building the Future. Proceedings of the VI International Rangeland Congress*. VI International Rangeland Congress Inc., Townsville, Australia, pp. 56–61.

Hassanein, N. and Kloppenburg, J.R. (1995) Where the grass grows again: knowledge exchange in the sustainable agriculture movement. *Rural Sociology* 60, 721–740.

Heywood, J., Hodgkinson, K., Marsden, S. and Pahl, L. (eds) (2000) *Graziers' Experiences in Managing Mulga Country*. Department of Primary Industries Queensland, Brisbane.

Holmes, J.H. (1996) Diversity and change in Australia's rangeland regions: translating resource values into regional benefits. *The Rangeland Journal* 19, 3–25.

Holmes, J.H. and Day, P. (1995) Identity, lifestyle and survival: value orientations of South Australian pastoralists. *The Rangeland Journal* 17, 193–212.

Kersten, S. and Ison, R. (1994) Diversity in yearly calendars on pastoral properties in western NSW: a constructivist perspective. *The Rangeland Journal* 16, 206–220.

Kloppenburg, J.R. (1991) Social theory and the de/reconstruction of agricultural science: local knowledge for an alternative agriculture. *Rural Sociology* 56, 519–548.

Lowe, P. (1957) *Jimmy and Pat meet the Queen*. Backroom Press, Broome, Australia.

Martin, S. (1993) *A New Land: European Perceptions of Australia, 1788–1850*. Allen and Unwin, Sydney.

Miller, D. (1987) *Material Culture and Mass Consumption*. Basil Blackwell, Oxford.

Murdoch, J. and Clark, J. (1994) Sustainable knowledge. *Geoforum* 25, 115–132.

National Rangeland Management Strategy (1996) National Management Working Group, NZECC/ARMAC, Canberra.

Piaget, J. (1951) *Play, Dreams and Imagination in Childhood*. Heinemann, London.

Piatelli-Palmarini, M. (1994) *Inevitable Illusions: How Mistakes of Misreason Rule our Minds*. John Wiley & Sons, New York.

Pretty, J.N. (1997) *Challenges in Agriculture and Rural Communities*. Proceedings of the Australian Pacific Extension Conference, Albury.

Pretty, J.N. and Shah, P. (1997) Making soil and water conservation sustainable: from coercion and control to partnerships and participation. *Land Degradation and Development* 8, 39–58.

Seddon, G. (1997) *Landprints, Reflections on Place and Landscape*. Cambridge University Press, Cambridge.

Stafford Smith, M. (1994) Sustainable production systems and natural resource management in the rangelands. *Outlook 94*, Australian Bureau of Agricultural and Resource Economics, Canberra, pp. 148–159.

Turnbull, D. (1997) Reframing science and other local knowledge traditions. *Futures* 29, 551–562.

Weeks, P. and Packard, J.M. (1997) Acceptance of scientific management by natural resource dependent communities. *Conservation Biology* 11, 236–245.

White, L. (1996) Are you an environmentalist or do you work for a living? Work and nature. In: Cronon, R. (ed.) *Uncommon Ground: Rethinking the Human Place in Nature*. W.W. Norton, New York.

Desertification and Soil Processes in Rangelands

5

David Tongway and Walter Whitford

Introduction

The term desertification was coined to graphically represent the state of the Sahelian lands in the 1970s when major drought accompanied by big increases in the human population served apparently to cause the desert margins to move into formerly more productive land (UN, 1977). The image of an encroaching desert is powerful and evocative and resulted in major international efforts to understand and deal with the problem. Since that time, the concept of desertification has been modified to the extent that the desert is no longer seen as inexorably increasing in size, nor restricted to the Sahel (Zuozhong and Xiangzhen, 1999; Arnalds, 2000). Most rangeland areas in the world have suffered some sort of degradation and recent reviews (Archer and Stokes, 2000) have shown the process to be not at all restricted to hot deserts or areas of high population density. This is not to deny, however, the major effects on the human populations using these lands and no doubt, much hardship has been endured.

Subsequently, attention has been directed towards assessing the loss of productive potential of the affected lands and understanding the biogeochemical implications of desertification. The temporal nature of loss of edaphic productivity has always been a key element of these studies. Is the loss of productivity long term? Can soil productivity be returned sooner rather than later? Was the observed effect 'superficial' and spontaneously reversible on the return of favourable seasons? What soil processes were affected by desertification and how can a better understanding of these improve our management of the macroscopic human problem in future?

This chapter provides a global perspective on the effects of desertification on soil processes. The human dimension of desertification has been covered previously (UN, 1977). In particular we ask whether approaches developed, processes identified and solutions elucidated in one part of the world can be applied in other regions without the need for additional primary research. If landscapes and soil processes can be understood at a sufficiently fundamental level, this systems knowledge can be adapted to other biomes in a relatively straight-forward manner. This would reduce the need for the difficult and expensive research on fine-scale biological processes in rangelands that are remote from laboratory facilities. This position contrasts with macroscopic studies where differences in human culture, species, climate, geology and so on conspire to make every locale unique. We also consider to what extent knowledge at fine or micro levels could provide insight at coarser scales and thus contribute to the development of monitoring systems that will identify critical thresholds or early warning signs. Is there a sufficient accumulation of knowledge that could be used to provide appropriate cost-effective rehabilitation of potential soil productivity?

Soil Processes and Desertification

In any biome, the physical, chemical and biological properties of the soil determine how stable it is to erosion, whether water is able to infiltrate into it, how much water it can store, and the rates and directions of nutrient cycling. These characteristics in turn determine soil productivity. In rangelands, the maintenance of soil processes and properties relies mostly on natural biological processes, in contrast to agricultural lands where management practices intervene to manipulate soil properties overtly and regularly. Herrick and Whitford (1999), discussing the continuum of processes from the 0.2 μm scale up to catchment scale, emphasize that at each scale there are processes mediated by different sets of biota, all contributing to the macroscopic behaviour of soil in terms of its stability and porosity (Fig. 5.1). In desertification, the central issue is the breakdown of processing of organic matter which provides both energy and nutrients to soil organisms. Most of the organic matter has its origin in vascular plants that capture atmospheric carbon. A hierarchy of soil organisms is involved in processing organic matter (Whitford and Herrick, 1995; Lavelle, 1997). The roles of these organisms are understood in general terms, but not in as much detail as many above-ground processes. Unanswered questions relate to the consequences flowing from the

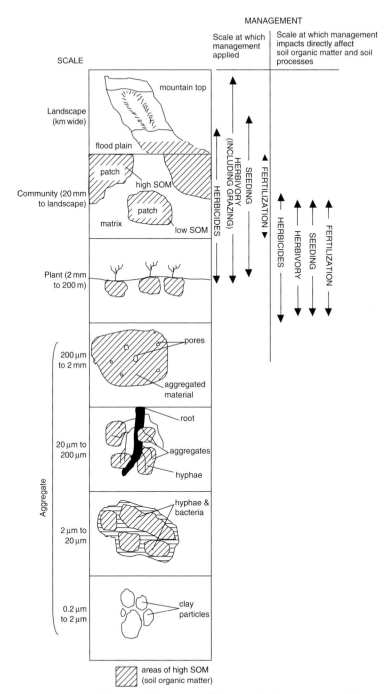

Fig. 5.1. Conceptual diagram illustrating differences in the scale at which management practices are applied and the scale at which they affect soil processes and properties such as organic matter cycling.

replacement of perennial plants with annuals or ephemerals in terms of altering the dynamics of the availability of organic matter and its quality in terms of mineralization potential. Whitford and Herrick (1995) have shown that profound changes can occur, and that these changes affect the ongoing functioning of the system. For example, there are changes in the below-ground carbon allocation between guilds of organisms; the nature of root exudates affects rhizosphere processes and mycorrhizal symbioses. *Knowledge of pathways involved in the transfer of organic compounds between plants and soil organisms is poor at a global level.*

To maintain key processes, a continual supply of 'fresh' organic matter is required (Oades, 1993). As desertification proceeds, the supply of organic matter to soil organisms is greatly attenuated, either because above-ground material is harvested by large herbivores or simply because production slows. The question of what happens to the soil-dwelling organic matter processing organisms during this hiatus is central to understanding the effect of desertification at fine scales. Kinnear and Tongway (1999) showed that some mite (*Acarina*) species, which have a central role in organic matter processing, were adversely affected by heavy grazing pressure, a precursor to desertification. Mites occupy and rely on very small air-filled pores in the soil. Their disappearance implies both lack of organic matter and partial collapse of soil structure. Soil compaction is often recorded as a consequence of desertification. This may operate at very fine scales where soil fauna provide both the mechanism and the structures that typify 'healthy' soils. Restitution dynamics appear positive in this case, but more study is necessary for a more complete understanding.

In rangelands, processes such as resource trapping by plants and other surface 'obstructions' during 'normal' times modify the immediate edaphic environment of vascular plants, improving their nutrient and water supplies to levels not predicted by macroscopic climatic summaries. Runoff/run-on and erosion/deposition processes result in distinct 'fertile patches' associated with perennial plants (Allsopp, 1999; Mazzarino and Bertiller, 1999; Northup and Brown, 1999a,b). These fertile patches support the production of above- and below-ground organic matter. Intuitively there should be a direct link between relatively easily observed surface features, such as plant density, through to elevated soil property levels and the micro-processing of organic matter. There is a strong feedback link from the latter to plant rhizosphere processes (Herrick and Whitford, 1999) and nitrogen cycling (Mazzarino and Bertiller, 1999) (Fig. 5.2).

Above-ground patchiness has been used by Tongway and Hindley (2000) in proposing an indicator system for monitoring desertification processes. They used the framework of Ludwig and Tongway (2000) which acknowledges the below-ground processes referred to above, at

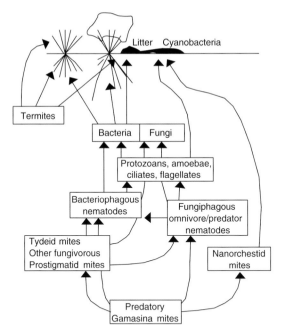

Fig. 5.2. Feeding relationships of soil fauna of a semi-arid ecosystem.

least in principle. Typically, the marked differences in soil properties at the plant/interplant scale have been referred to as 'heterogeneity'. This was necessary to show how different natural landscapes are from managed agricultural landscapes that are homogenized by ploughing, fertilizer and water additions. However, it is not the heterogeneity *per se* that is important here, but the fact that the fertile patches had soil property values well above critical thresholds for the maintenance of perennial plants in adverse climates. The loss of these plants and their fertile patches (Tongway, 1991) triggers the soil degradation or desertification process: that is, the edaphic environment falls below a critical threshold and there are few autogenous processes for their reconstitution. Feedback mechanisms close down. This has been widely reported (Belnap *et al.*, 1999; Eldridge *et al.*, 1999; Zuozhong and Xiangzhen, 1999).

Ludwig and Tongway (2000) have suggested a 'systems' approach to the assessment of landscape function. They advocated an objective description of the loss of soil productive potential in place of the emotive term of desertification. They proposed a conceptual structure or framework to organize knowledge and make provision for cross-scale analysis, using information such as that referred to above.

This framework uses processes and their interactions as its basic input information set, expressly recognizing spatial effects and sequences of processes. It is possible to construct various simulation modules as part of this framework, representing runoff/run-on or nutrient cycling processes and to take into account consumptive processes such as fire and grazing.

Future Directions

Knowledge of within-soil processes is minimal for our purposes, but by using a framework that expressly covers a broad functional scale and integrates up to patch and local catchment scales, workable monitoring procedures for assessing soil productive potential are possible. Greater knowledge of the biology of soil processes is required, particularly in relation to alternative pathways and 'bottlenecks', the capacity of suites of organisms to survive in desertified circumstances and the consequences if they do not. Recolonization pathways for key groups of organism must be better known at both scientific and management levels to ensure that the processes they mediate can be restarted.

We propose Fig. 5.3 as exemplifying the activity in the soil processes area. If these activities are kept in balance and in mutual communication, the technical tools necessary to deal with desertification are within our grasp.

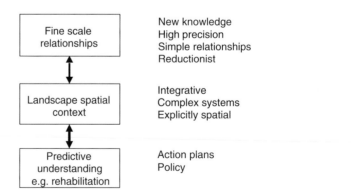

Fig. 5.3. Interactions between basic research, integrated landscape frameworks and policy instruments needed to deal with desertification.

References

Allsopp, N. (1999) Consequences to soil patterns in Southern Africa following vegetation change with livestock management. In: Eldridge, D. and Freudenberger, D. (eds) *People of the Rangelands. Building the Future. Proceedings of the VI International Rangeland Congress*. VI International Rangeland Congress Inc., Townsville, Australia, pp. 96–101.

Archer, S. and Stokes, C. (2000) Stress, disturbance and change in rangeland ecosystems. In: Arnalds, O. and Archer, S. (eds) *Rangeland Desertification*. Kluwer Academic, Dordrecht, pp. 17–38.

Arnalds, O. (2000) Desertification: an appeal for a broader perspective. In: Arnalds, O. and Archer, S. (eds) *Rangeland Desertification*. Kluwer Academic, Dordrecht, pp. 5–15.

Belnap, J., Ojima, D., Philips, S. and Barger, N. (1999) Biological soil crusts of Mongolia: impacts of grazing and precipitation on nitrogen inputs. In: Eldridge, D. and Freudenberger, D. (eds) *People of the Rangelands. Building the Future. Proceedings of the VI International Rangeland Congress*. VI International Rangeland Congress Inc., Townsville, Australia, pp. 130–131.

Eldridge, D., Zaady, E., Shakak, M. and Myers, C. (1999) Control of desertification by microphytic crusts in a Negev Desert shrubland. In: Eldridge, D. and Freudenberger, D. (eds) *People of the Rangelands. Building the Future. Proceedings of the VI International Rangeland Congress*. VI International Rangeland Congress Inc., Townsville, Australia, pp. 111–113.

Herrick, J.E. and Whitford, W.G. (1999) Integrating soil processes into management: from microaggregates to macrocatchments. In: Eldridge, D. and Freudenberger, D. (eds) *People of the Rangelands. Building the Future. Proceedings of the VI International Rangeland Congress*. VI International Rangeland Congress Inc., Townsville, Australia, pp. 91–95.

Kinnear, A. and Tongway, D.J. (1999) Responses of mite communities to rangeland degradation and grazing. In: Eldridge, D. and Freudenberger, D. (eds) *People of the Rangelands. Building the Future. Proceedings of the VI International Rangeland Congress*. VI International Rangeland Congress Inc., Townsville, Australia, pp. 118–119.

Lavelle, P. (1997) Faunal activities and soil processes: adaptive strategies that determine ecosystem function. *Advances in Ecological Research*. Academic Press, New York, pp. 93–132.

Ludwig, J.A. and Tongway, D.J. (2000) Viewing rangelands as landscape systems. In: Arnalds, O. and Archer, S. (eds) *Rangeland Desertification*. Kluwer Academic, Dordrecht, pp. 39–52.

Mazzarino, M.J. and Bertiller, M.B. (1999) Soil N pools and processes as indicators of desertification in semi-arid woodlands and semi-arid to arid steppes of Argentina. In: Eldridge, D. and Freudenberger, D. (eds) *People of the Rangelands. Building the Future. Proceedings of the VI International Rangeland Congress*. VI International Rangeland Congress Inc., Townsville, Australia, pp. 101–105.

Northup, B.K. and Brown, J.R. (1999a) Spatial distribution of soil carbon in grazed woodlands of dry tropical Australia: tussock and inter-tussock scales. In: Eldridge, D. and Freudenberger, D. (eds) *People of the Rangelands. Building the Future. Proceedings of the VI International Rangeland Congress*. VI International Rangeland Congress Inc., Townsville, Australia, pp. 120–121.

Northup, B.K. and Brown, J.R. (1999b) Spatial distribution of soil carbon in grazed woodlands of dry tropical Australia: meso-patch to community scales. In: Eldridge, D. and Freudenberger, D. (eds) *People of the Rangelands. Building the Future. Proceedings of the VI International Rangeland Congress.* VI International Rangeland Congress Inc., Townsville, Australia, pp. 121–122.

Oades, J.M. (1993) The role of biology in the formation, stabilization and degradation of soil structure. *Geoderma* 56, 377–400.

Tongway, D.J. (1991) Functional analysis of degraded rangelands as a means of defining appropriate restoration techniques. In: Gaston, A., Kerrick, M. and Le Houerou, H. (eds) *Proceedings of the Fourth International Rangeland Congress.* Association Française de Patoralisme, Montpellier, France, pp. 166–168.

Tongway, D.J. and Hindley, N. (2000) Assessing and monitoring desertification with soil indicators. In: Arnalds, O. and Archer, S. (eds) *Rangeland Desertification.* Kluwer Academic, Dordrecht, pp. 89–98.

UN (1977) *Desertification, its Causes and Consequences.* Pergamon Press, Oxford.

Whitford, W.G. and Herrick, J.E. (1995) Maintaining soil processes for plant productivity and community dynamics. In: *Rangelands in a Sustainable Biosphere: Proceedings of the Fifth International Rangeland Congress.* Society for Range Management, Denver, Colorado, pp. 33–37.

Zuozhong, C. and Xiangzhen, L. (1999) Degradation of the grassland ecosystem in China and its biological processes. In: Eldridge, D. and Freudenberger, D. (eds) *People of the Rangelands. Building the Future. Proceedings of the VI International Rangeland Congress.* VI International Rangeland Congress Inc., Townsville, Australia, pp. 105–107.

Understanding and Managing Rangeland Plant Communities

6

Steve Archer and Alison Bowman

Introduction

An understanding of plant communities and their dynamics is central to management aimed at minimizing degradation, promoting restoration or sustaining productivity of the world's rangelands. The spatial organization and temporal dynamics of communities are influenced by resource availability (e.g. water, nutrients), stresses (e.g. temperature, salinity) and disturbances (e.g. fire, grazing) as these affect plant performance. The differential responses of plants variously adapted to acquire resources and tolerate stress and disturbance affects species interactions and population dynamics (recruitment, longevity and mortality). Resource availability, stress and disturbance also vary with time and across space. Soils and topography modulate plant and community responses to these. This spatial and temporal variation produces patterns and 'behaviours' in communities and may induce fluctuation or directional change in community composition.

The challenge facing ecologists and managers is to recognize and understand the constraints imposed by these factors at various spatial and temporal scales and determine how and when they might be effectively manipulated or modified to reach desired goals. There are various approaches to achieving this recognition and understanding. This chapter will argue that: *(i) the role of nitrogen as a determinant of plant communities in rangelands has been underestimated; (ii) the importance of positive species interactions (facilitation) has been under-appreciated; and (iii) the benefits that might accrue from explicitly combining descriptive, experimental, monitoring and modelling approaches in a hierarchical framework have yet to be realized.*

Contrasting Perspectives on Community Organization

Limiting factors: water vs. nitrogen?

Low and variable annual rainfall is a prominent feature of many rangelands. Moisture has typically been regarded as *the* limiting resource and driving force in community dynamics. However, nutrient availability may also exert a strong influence. There are clear evolutionary trade-offs between features enabling plants to tolerate nutrient-poor conditions and features conferring competitive superiority under nutrient-rich conditions (Chapin, 1980, 1993; Berendse and Elberse, 1990; Aerts and van der Peijl, 1993). Furthermore, plants can modify soil nutrient status (Hobbie, 1992) in ways which may promote or deter community change (Tilman and Wedin, 1991; Binkley and Giardina, 1998; Schlesinger and Pilmanis, 1998). Linkages between nutrient cycling and plant community dynamics may thus be strong (Rietkerk and van de Koppel, 1997; Wedin, 1999). *To what extent has our focus on water in isolation from nutrients constrained our understanding and management of plant communities?*

It is generally assumed that at lower levels of annual precipitation, above-ground net primary productivity is limited primarily by water, whereas at higher levels of precipitation, it is limited primarily by nitrogen. Hooper and Johnson (1999) tested this assumption by synthesizing results from fertilization experiments in arid, semi-arid and subhumid rangelands. Their survey found no strong evidence of a shift from a water to a nutrient limitation across a wide geographic rainfall gradient. Indeed, responses to N addition were typically positive, even at dry locations and even in years of below average rainfall. Such results suggest tight coupling between water and nitrogen and co-limitation (Chapin *et al.*, 1987; Chapin, 1991), an interpretation also supported by process-based dynamic simulation models (Schimel *et al.*, 1997).

Plant community studies that focus solely on water without accounting for plant-available soil nitrogen may be overlooking a critical factor. Contradictions in predictions of plant community response to moisture might be resolved if nitrogen is factored in. The physiological and evolutionary responses of plants to nutrient limitation and the responses of microbial decomposers to plant tissue chemistry create feedbacks that may reinforce N limitations

(Vitousek, 1982; Hobbie, 1992; Chapin, 1993). Disturbances such as grazing and fire may alter or disrupt the feedbacks between vegetation and N availability (Fig. 6.1) (Holland *et al.*, 1992; Seastedt, 1995; Wedin, 1995, 1999) and propel a community into alternate stable states (Jefferies *et al.*, 1994; Pastor and Cohen, 1997; Rietkerk and van de Koppel, 1997; Rietkerk *et al.*, 1997).

All temperate and tropical biomes receive more N via wet and dry deposition today than pre-industrially; and northern hemisphere temperate ecosystems receive more than four times that of pre-industrial levels (Holland *et al.*, 1999). Given these recent increases in N deposition, there is a pressing need to understand how water and N influence ecosystem processes both independently and interactively (Burke *et al.*, 1991; Vitousek *et al.*, 1997). If, for example, N deposition reduces or alleviates N limitations in rangelands, primary production and species composition may become more sensitive to temporal variation in rainfall and change the nature of management risk and uncertainty.

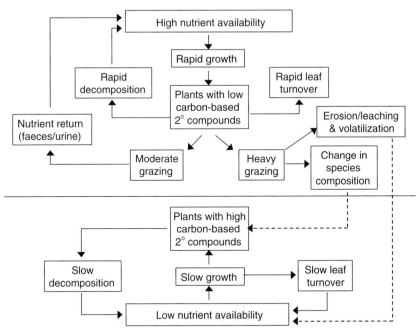

Fig. 6.1. Conceptual model of plant–soil feedback in low (upper panel) and high (lower panel) fertility sites and how prolonged heavy grazing might transform a high fertility site to a low fertility site, by altering species composition and plant–soil interactions (adapted from Chapin, 1993). In this conceptual model, nutrients, rather than water, drive community response to grazing.

Species interactions: competition vs. facilitation?

Established plants exert a sphere of influence on soils and micro-climate in the vicinity of their canopies. This sphere of influence has typically been viewed from the perspective of competition (Keddy, 1989; Walker *et al.*, 1989). However, plants may also serve as recruit-ment foci and create conditions conducive to the germination, establishment or growth of other plants. As such, *positive interactions among species (facilitation) may play an important, but under-appreciated role in the organization and dynamics of plant communi-ties.* Under what conditions is facilitation likely to occur and to what extent has the focus on competition rather than facilitation constrained our understanding and management of plant communities?

Bertness and Callaway (1994) and Callaway (1995) persuasively argue that evidence for the importance of facilitation in community organization and dynamics has accrued to the point where it warrants formal inclusion into community ecological theory. Plants may facilitate other plants directly or actively by ameliorating harsh environmental conditions, by altering soil properties or by increasing availability of resources. Facilitation may be indirect or passive if a plant eliminates competitors, introduces or attracts other beneficial organisms (e.g. microbes, pollinators), provides protection from herbivory, or serves as a focus for the concentration of propagules.

Positive interactions are prominent in some communities and conspicuously absent in others. It appears that their relative importance varies with species traits (Callaway, 1998a) and changes with time and the life stages of the interacting plants (e.g. Greenlee and Callaway, 1996; Barnes and Archer, 1998) or with features such as plant density (cf. Scholes and Archer, 1997). There may also be variation among individuals within a community (Callaway and Tyler, 1999). Generalizations regarding facilitation, as with those regarding competition, should therefore be made cautiously.

Interactions among plants have been shown to shift from comp-etition to facilitation along environmental continua (Archer, 1995; Callaway, 1998b). Facilitation may be most common in communities developing under high physical stress and in communities with high consumer pressure (Bertness and Callaway, 1994; Callaway and Walker, 1997). In these situations, amelioration of stress by neighbours may enhance growth more than competition restricts it. In intermediate habitats, where the physical environment is relatively benign and consumer pressure is less severe, rapid resource acquisition is possible and competitive interactions may be a dominant structuring force. Incorporation of facilitation into models of community organiza-tion that are largely dominated by competition, lottery events and fluctuations in stress and resource availability, may pave the way to clearer understanding (Fig. 6.2).

Main effects vs. interactions

Ecologists and range managers tend to view categories of variables associated with resource availability, stress and disturbance as independent 'main effects'. Consider the vast number of papers focused on the role of precipitation, the role of grazing, the role of fire, the role of

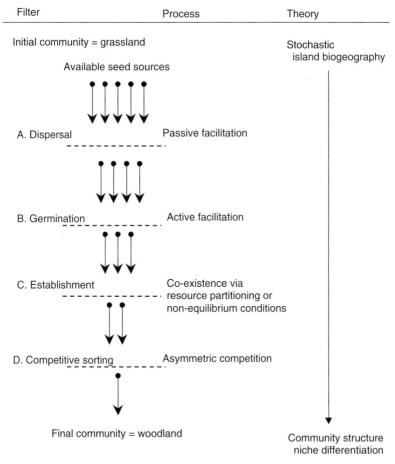

Fig. 6.2. Conceptual model of species–environment interactions during succession from grassland to woodland demonstrating the rich array of processes and interactions that interact to affect community structure and change (based on Archer, 1995 and Stokes, 1999).

diversity, etc. in shaping plant communities. Typically, these factors are considered independently and in isolation from each other when, in fact, they are highly interactive. For example, the effects of grazing may be minimized in years of good rainfall and intensified in years of low rainfall. Grazing and rainfall will also affect fine fuel biomass and continuity and thereby fire frequency and intensity. Thus, a realistic understanding of the effects of grazing, fire or precipitation on community structure and function is contingent upon understanding their interactions.

Major funding programmes over the decades have changed the emphasis of research, yet the tendency to focus on 'main effects' persists. In the 1960s–1970s there was a focus on abiotic (climatic) factors with little emphasis on biotic effects on ecosystem structure and function (i.e. the International Biological Programme). In the 1970s–1980s, there was widespread recognition of the role of animals in affecting plant communities; however, plant–animal interaction studies were often conducted with little regard for abiotic influences (e.g. climate, nutrients and fire). To what extent has the focus on 'main effects' constrained our understanding and managing of plant communities?

Field experiments are usually restricted to examining a limited subset of possible effects, to the exclusion of dominant interactive effects. Results of field experiments are therefore highly context-dependent. O'Connor (1999) illustrates the context problem using a series of separate, long-term factorial experiments initiated in 1948, that were designed to investigate the effect of nutrients, fire, mowing and rotational livestock grazing on a grassland community. Each of the experiments clearly demonstrated that resource availability and type of disturbance had significant effects on community composition. However, despite the impressive, long-term nature of this experiment the relative importance of these independent factors remains open to debate as does the question of 'How would the community have responded if some or all of these factors had interacted?' In the absence of explicit theoretical predictions to guide experimentation, there is a preoccupation with simply demonstrating that a factor is 'important'. Preoccupation with demonstrating that specific factors are important will produce catalogues of examples. These in turn foster analyses of whether an observed phenomenon is caused by this or that factor or the relative importance of selected factors. In interacting systems, this may be fruitless enterprise as it is conceptually impossible to assign quantitative values to specific causal factors or separate them in this way (Levins and Lewontin, 1985). The emphasis should instead be on how factors interact and the nature of their interconnectedness. We should not be searching for factors *per se*. Rather, *we should be endeavouring to construct coherent conceptual frameworks for predicting the consequences of factor interactions.*

Description vs. experimentation

Early studies of communities were primarily descriptive and quantified *how* communities looked or *how* they changed. Processes were inferred from patterns and space was substituted for time as a means for assessing community change. However, inferences from descriptive studies can be misleading (Austin, 1977; Shugart *et al.*, 1981; Likens, 1988; Cale *et al.*, 1989). Furthermore, descriptive studies often lack explanations of *why* observed community changes occur. Long-term observations can suggest importance of exogenous events (such as drought or a late freeze) on communities, without revealing how endogenous processes were modified to produce the observed response.

The following example illustrates the pitfalls of making inferences from descriptive studies. Field observations of plant distributions and soil properties in a savanna parkland landscape demonstrated that large groves of woody vegetation occurred where argillic horizons (zones of clay accumulation) were poorly expressed. Where the argillic horizon was well developed, small shrubs and grasses dominated (Archer, 1995). Soil trenches revealed that burrowing rodent and leaf cutter ant activity was substantial in soils with the poorly developed argillic horizons associated with tree groves and minimal in non-grove soils. These observations led to the 'explanation' that mixing of soils by cutter ants and burrowing rodents had disrupted a laterally continuous argillic horizon and hence promoted the development of tree groves. This explanation was logical, intuitively appealing and consistent with field data. Using the method of multiple working hypotheses (Chamberlin, 1965), it was reasoned that if this explanation were valid, the clay content of grove and non-grove soils should be comparable when summed across the entire soil profile. As it turned out, this was not the case. It therefore appears that woody plants, cutter ants and burrowing rodents were exploiting a pre-existing condition on the landscape, where for pedogenic reasons, the argillic horizon had never formed. Subsequent research has supported the latter explanation (Boutton, 1996; Stroh *et al.*, 2001).

Reductionist approaches based on experimentation and manipulation have been advocated as an alternative to descriptive approaches. The hypothesis testing approach seeks to answer the *why* question via rigorous application of the scientific method and to avoid the pitfalls exemplified in the preceding example. However, to control and manipulate the environment typically necessitates working on small scales and over short time frames. Surveys of experimental studies reveal that about 50% have been conducted in plots 1 m² and 40% have been completed in 1 year or less (Kareiva and Anderson, 1988; Brown and Roughgarden, 1989; Tilman, 1989). Do such studies really advance our understanding of dynamic, complex communities?

Field experiments: problems and pitfalls

> . . . there is no single, simple approach that can ever unambiguously demonstrate how or why a particular process, physical factor, or species has an effect on another element of the ecosystem . . . ecological research requires a synthetic approach in which observation, experimental, and theoretical approaches are pursued in a simultaneous, coordinated, interactive manner.
>
> (Tilman, 1989, p. 136)

We tend to measure things for which we have tools and we assume that what we measure is important. There is also a tendency to avoid rather than include stochasticity, biocomplexity and variability. Field experiments are often too short in duration, too small in spatial scale and too narrowly focused to effectively capture characteristic behaviours of communities (e.g. Watson *et al.*, 1996). Additional processes, undetected or not represented at the scale of the experiment, may dictate the structure and dynamics of communities at spatial and temporal scales relevant to management (Turner and Dale, 1998). *Field experiments may therefore be highly contextual, with artificialities that make their extrapolation in time and space tenuous* (Bender *et al.*, 1984; Diamond, 1986; Yodzis, 1988; Inchausti, 1994).

For example, experiments whose results support the notion that plant species diversity enhances ecosystem productivity and resilience (Tilman and Downing, 1994; Tilman *et al.*, 1996) have been challenged on the basis that these traits are determined largely by the most productive species in the experimental plots, irrespective of plot diversity (Aarssen, 1997; Grime, 1997). The correlation between diversity and community productivity and resilience may simply reflect the fact that the most productive species used in the study had a greater chance of being included in the more diverse plots than in the less diverse plots (Huston, 1997).

Biological variability and complex organismic interactions should be included rather than avoided in experiments, even if the price to be paid is a less clear-cut mechanistic insight. Reductionist, highly controlled experiments may contain 'hidden treatments' (Huston, 1997) and exclude or limit effects critically affecting community dynamics. Large-scale, long-term experiments focused on factor interactions, even if expensive and messy, are needed if we are to understand plant communities at scales relevant to management and socio-economic policy. Further, there needs to be an explicit integration of experimentation with theory. In the absence of explicit theoretical predictions to guide experimentation, we end up with catalogues of important but disconnected variables. Experiments should be harnessed to adjudicate theory or major conceptual frameworks or to measure quantities that can be employed with the theory to make more specific predictions for further tests (Werner, 1998). Proliferation of small-scale, short-term experiments divorced from theory will contribute information and data to a body of knowledge, but may do little to advance our understanding.

Hierarchical perspectives

The questions before us are not whether we should do experiments in community ecology or to what extent. We should, and in abundance. Nor is the question whether experiments are the only way to contribute toward a predictive ecology. They are not.

(Werner, 1998, p. 3)

Plant communities comprise myriad interacting and interdependent elements. How do we simplify their daunting complexity to manageable proportions? Hierarchy theory is one approach (Allen and Starr, 1982; O'Neill *et al.*, 1986; Rosswall *et al.*, 1988). In this conceptual view, ecological systems are represented as a graded series with several levels of organization. An entity representing a given level of organization consists of smaller entities and is a component of a higher level of the hierarchy (Fig. 6.3).

For example, an individual plant comprises interacting leaf, stem and root subsystems. However, this same plant, if rooted in a soil along with other plants, is a component of a higher-level entity, which might be recognized as a patch. Patches arrayed across a soil type may collectively represent a community; communities are distributed along catenas to form landscapes, etc.

As this example implies, there are distinctions between structural entities at a given level of organization (e.g. between roots, stems and leaves at the plant level; between plants, animals and microbes at the patch level); and distinctions between successive levels (between leaves, plants, patches, communities). Each level of organization is characterized by processes that operate at certain spatial and temporal scales. Plant level processes would typically focus on gas exchange, water relations and allocation. Patch level processes might focus on infiltration rates, seedling establishment, competitive interactions, and herbivore forage selection. At the community level, distinctions between individual plants are lost, but runoff–runon, dry deposition,

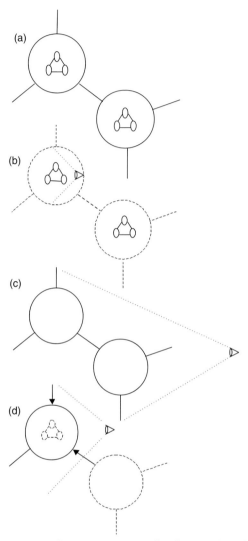

Fig. 6.3.   The perspective taken on a system will influence the information accessible at various levels of organization. This hypothetical system consists of two entities, each with three parts (a). The complete system is not visible within any single observation set. Inside the surface, looking inward (b) is the only position from which the parts and their interconnections can be seen without distortion. If the observer moves far enough away from the surface, the other whole is identifiable as a separate entity, responsible for part of the environmental influence (c). Seen from outside, the parts are obscured by the intervening surface and the other entity is manifested only as an environmental influence of undefined origin (d). The eye indicates the position from which the system is observed in each case (after Allen *et al.*, 1984, 1999).

diversity, boundary dynamics and edge effects are now recognized. Thus, higher levels in an ecological hierarchy contain, constrain, behave at lower frequencies and exhibit less bond strength than lower levels. In addition, higher levels buffer lower levels and filter environmental influences and variability (Allen *et al.*, 1984, 1999). Therefore, unexplained variance or behaviour at lower levels might be accounted for when higher order effects are explicitly acknowledged.

As with the parable of the blind men who each felt a different part of an elephant and proceeded to describe the whole without knowledge of the other parts, our perception and understanding of communities may be largely a matter of perspective. *In contrast to reductionist approaches, the hierarchical approach permits evaluation of complex systems without reducing them to a series of simple, disconnected components. No single level in an ecological hierarchy is fundamental; understanding a system at one level requires knowledge of levels both above and below the targeted level.* Interpretation of system behaviour at one level of organization without consideration of adjacent levels is therefore out of context. For example, the views of Clements, Gleason and Tansley may be more complementary than contradictory when viewed from a hierarchical perspectives (Fig. 6.4) (Hoekstra *et al.*, 1991).

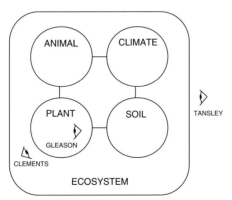

Fig. 6.4. A schematic representation of the different scales of perception involved in the individualistic concept of community (Gleason, 1926), viewed from inside the community, where the focus is on the autonomy of the component species; the superorganismal concept of community (Clements, 1905), viewed from outside the community so as to emphasize its integrity; and Tansley's (1935) conception of the ecosystem, viewing the system from a greater distance, so that the autonomy of the biota is obscured as it is integrated with the physical environment. The three images of an eye represent the locations of the observers (from Hoekstra *et al.*, 1991).

Holistic and reductionist approaches should not be viewed as mutually exclusive. Each provides a unique perspective. The reductionist approach dissects lower levels of organization and provides mechanistic explanations and insights into how systems work. However, reductionist studies strictly looking 'inside' the system do not see the whole and its emergent properties. The holistic approach views a system in the context of the higher levels in which it is embedded, and provides insight into the significance of phenomena at lower levels. The search for mechanisms should therefore be balanced by concern for significance (Passioura, 1979; Lidicker, 1988). Studies focused at one level of organization without regard for higher levels can thus generate vast amounts of information, but little understanding.

The Way Forward

We have a wealth of detailed observations on the natural history of our planet, but are only beginning to uncover (or invent) the general principles which can organize this mass of observations.

(Keddy, 1989)

We are drowning in information, while starving for knowledge. The world henceforth will be run by synthesisers, people able to put together the right information at the right time, think critically about it, and make important choices wisely.

(Wilson, 1998)

The above quotes indicate that what is needed is more understanding and new perspectives, not simply more data. Community composition and dynamics are outcomes of the interactions among constellations of driving variables. Therefore, extrapolations from context-dependent experiments and descriptive studies should be made with caution, and static management 'prescriptions' based on case studies should be viewed with scepticism. How do we then progress with understanding and managing plant communities?

The preoccupation with 'main effects' is partially due to logistical constraints: the duration of contract/grant funding typically relegates most studies to short (2–3 year) time frames and a very specific, narrow focus. A clear articulation of the need for more comprehensive studies designed to focus on key interactions may be a necessary first step in overcoming this logistical barrier. *Descriptive, experimental and modelling approaches have advantages and disadvantages, each providing perspectives the others cannot. Natural resource administrators and science programme managers should therefore promote multidisciplinary ventures that proactively integrate these approaches* (Fig. 6.5).

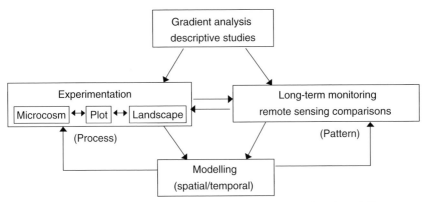

Fig. 6.5.   Conceptual integration of descriptive, experimental and modelling approaches. To date, most of these approaches have been used in isolation. Gradient analysis may suggest hypotheses, which could be tested via experimentation and monitoring. These, in turn, may suggest new suites of environmental variables for gradient analysis while providing input for simulation models. Simulation models and modelling experiments feed back to help prioritize and refine experiments and monitoring protocol. Linked remote sensing–modelling approaches hold the promise to provide monitoring of function as well as structure over large areas (e.g. Asner *et al.*, 1998; Wessman, 1992; Wessman *et al.*, 1997). Experimentation and monitoring should be conducted at spatial and temporal scales appropriate to specified levels of hierarchical organization (Fig. 6.3). Experiments should be harnessed to adjudicate theory or major conceptual frameworks or to measure quantities that can be employed with the theory to make more specific predictions for further tests.

The 'multiple working hypotheses' approach (Chamberlin, 1965; Ward, 1993) has clear utility in community ecology, yet remains under-utilized. Astronomers, geologists, climatologists and ocean-ographers have achieved marked successes in inferring process from pattern, in constructing and evaluating complex models, and in testing hypotheses without the benefit of experimental manipulation and replication (Brown, 1994). We must move beyond our traditional, simplistic 'either–or' mentality (either water or nitrogen as *the* limiting factor; either competition or facilitation as *the* driver of species inter-actions; either descriptor or experimentation or modelling as *the* approach to studying communities). Perspectives which embrace the duality of resource constraints (e.g. water *and* nitrogen) and processes (e.g. competition *and* facilitation) as determinants of plant communi-ties and which integrate complementary approaches for studying these (e.g. experimentation *and* description *and* modelling as guided by theory) are likely to provide us with a richer, more robust understanding of plant communities and ecosystems.

References

Aarssen, L.W. (1997) High productivity in grassland ecosystems: effected by species diversity or productive species? *Oikos* 80, 183–184.

Aerts, R. and van der Peijl, M.J. (1993) A simple model to explain the dominance of low-productive perennials in nutrient-poor habitats. *Oikos* 66, 144–147.

Allen, T.F.H. and Starr, T.B. (1982) *Hierarchy: Perspectives for Ecological Complexity.* University of Chicago Press, Chicago.

Allen, T.F.H., O'Neill, R.V. and Hoekstra, T.W. (1984) Interlevel relations in ecological research and management: some working principles from hierarchy theory. General Technical Report RM-110, USDA Forest Service, Fort Collins, Colorado.

Allen, T.F.H., O'Neill, R.V. and Hoekstra, T.W. (1999) Interlevel relations in ecological research and management: some working principles from hierarchy theory. In: Dodson, S.I., Allen, T.F.H., Carpenter, S.R., Elliot, K., Ives, A.R., Jeanne, R.L., Kitchell, J.F., Langston, N.E. and Turner, M.G. (eds) *Readings in Ecology.* Oxford University Press, New York, pp. 393–412.

Archer, S. (1995) Tree–grass dynamics in a *Prosopis*–thornscrub savanna parkland: reconstructing the past and predicting the future. *Ecoscience* 2, 83–99.

Asner, G.P., Bateson, C.A., Privette, J.L., Elsaleous, N. and Wessman, C.A. (1998) Estimating vegetation structural effects on carbon uptake using satellite data fusion and inverse modeling. *Journal of Geophysical Research-Atmospheres* 103, 28839–28853.

Austin, M.P. (1977) Use of ordination and other multivariate descriptive methods to study succession. *Vegetatio* 35, 165–175.

Barnes, P.W. and Archer, S.R. (1998) Tree–shrub interactions in a subtropical savanna parkland: competition or facilitation? *Journal of Vegetation Science* 10, 525–536.

Bender, E.A., Case, T.J. and Gilpin, M.E. (1984) Perturbation experiments in community ecology: theory and practice. *Ecology* 65, 1–13.

Berendse, F. and Elberse, W.T. (1990) Competition and nutrient availability in heathland and grassland communities. In: Grace, J.B. and Tilman, D. (eds) *Perspectives on Plant Competition.* Academic Press, San Diego, pp. 93–116.

Bertness, M.D. and Callaway, R. (1994) Positive interactions in communities. *Trends in Ecology and Evolution* 9, 191–193.

Binkley, D. and Giardina, C. (1998) Why do tree species affect soils? The warp and woof of tree–soil interactions. *Biogeochemistry* 42, 89–106.

Boutton, T.W. (1996) Stable carbon isotope ratios of soil organic matter and their use as indicators of vegetation and climate change. In: Boutton, T.W. and Yamasaki, S.I. (eds) *Mass Spectrometry of Soils.* Marcel Dekker, New York, pp. 47–82.

Brown, J.H. (1994) Grand challenges in scaling up environmental research. In: Michener, W.K., Brunt, J.W. and Stafford, S.G. (eds) *Environmental Information Management and Analysis: Ecosystem to Global Scales.* Taylor and Francis, London, pp. 21–26.

Brown, J.H. and Roughgarden, J. (1989) US ecologists address global change. *Trends in Ecology and Evolution* 4, 255–256.

Burke, I.C., Kittel, T.G.F., Lauenroth, W.K., Snook, P., Yonker, C.M. and Parton, W.J. (1991) Regional analysis of the central Great Plains: sensitivity to climate variability. *Bioscience* 41, 685–692.

Cale, W.G., Henebry, G.M. and Yeakley, J.A. (1989) Inferring process from pattern in natural communities: can we understand what we see? *Bioscience* 39, 600–605.

Callaway, R.M. (1995) Positive interactions among plants. *Botanical Review* 61, 306–349.

Callaway, R.M. (1998a) Are positive interactions species-specific? *Oikos* 82, 202–207.

Callaway, R.M. (1998b) Competition and facilitation on elevation gradients in subalpine forests of the northern Rocky Mountains, USA. *Oikos* 82, 561–573.

Callaway, R.M. and Tyler, C. (1999) Facilitation in rangelands: direct and indirect effects. In: Eldridge, D. and Freudenberger, D. (eds) *People of the Rangelands. Building the Future. Proceedings of the VI International Rangeland Congress.* VI International Rangeland Congress Inc., Townsville, Australia, pp. 197–202.

Callaway, R.M. and Walker, L.R. (1997) Competition and facilitation: a synthetic approach to interactions in plant communities. *Ecology* 78, 1958–1965.

Chamberlin, T.C. (1965) The method of multiple working hypotheses. *Science* 148, 754–759.

Chapin, F.S., III (1980) The mineral nutrition of wild plants. *Annual Review of Ecology and Systematics* 11, 233–260.

Chapin, F.S., III (1991) Effects of multiple environmental stresses on nutrient availability and use by plants. In: Mooney, H.A., Winner, W.E. and Pell, E.J. (eds) *Response of Plants to Multiple Stresses.* Academic Press, New York, pp. 67–88.

Chapin, F.S., III (1993) Functional role of growth forms in ecosystem and global processes. In: Ehleringer, J.R. and Field, C.B. (eds) *Scaling Physiological Processes: Leaf to Globe.* Academic Press Inc., San Diego, California, pp. 287–311.

Chapin, F.S., III, Bloom, A.J., Field, C.B. and Waring, R.H. (1987) Plant responses to multiple environmental factors. *Bioscience* 37, 49–57.

Clements, F.E. (1905) *Research Methods in Ecology.* Arna Press, New York.

Diamond, J. (1986) Overview: laboratory experiments, field experiments and natural experiments. In: Diamond, J. and Case, T.J. (eds) *Community Ecology.* Harper and Row, New York, pp. 3–22.

Gleason, H.A. (1926) The individualistic concept of the plant association. *Bulletin of the Torrey Botanical Club* 53, 1–20.

Greenlee, J.T. and Callaway, R.M. (1996) Abiotic stress and the relative importance of interference and facilitation in montane bunchgrass communities in western Montana. *American Naturalist* 148, 386–396.

Grime, J.P. (1997) Biodiversity and ecosystem function: the debate deepens. *Science* 277, 1260–1261.

Hobbie, S.E. (1992) Effects of plant species on nutrient cycling. *Trends in Ecology and Evolution* 7, 336–339.

Hoekstra, T.W., Allen, T.F.H. and Flather, C.H. (1991) Implicit scaling in ecological research. *Bioscience* 41, 148–154.

Holland, E.A., Parton, W.J., Detling, J.K. and Coppock, D.L. (1992) Physiological responses of plant populations to herbivory and their consequences for ecosystem nutrient flow. *American Naturalist* 140, 685–706.

Holland, E.A., Dentener, F.J., Braswell, B.H. and Sulzman, J.M. (1999) Contemporary and pre-industrial global reactive nitrogen budgets. *Biogeochemistry* 46, 7–43.

Hooper, D.U. and Johnson, L. (1999) Nitrogen limitation in dryland ecosystems: responses to geographical and temporal variation in precipitation. *Biogeochemistry* 46, 247–293.

Huston, M.A. (1997) Hidden treatments in ecological experiments: re-evaluating the ecosystem function of biodiversity. *Oecologia* 110, 449–460.

Inchausti, P. (1994) Reductionist approaches in community ecology. *American Naturalist* 143, 201–221.

Jefferies, R.L., Klein, D.R. and Shaver, G.R. (1994) Vertebrate herbivores and northern plant communities: reciprocal influences and responses. *Oikos* 71, 193–206.

Kareiva, P. and Anderson, M. (1988) Spatial aspects of species interactions: the wedding of models and experiments. In: Hastings, A. (ed.) *Community Ecology*. Springer, New York, pp. 38–54.

Keddy, P.A. (1989) *Competition*. Chapman and Hall, New York.

Levins, S. and Lewontin, R. (1985) *The Dialectical Biologist*. Harvard University Press, Cambridge, Massachusetts.

Lidicker, W.Z. (1988) The synergistic effects of reductionist and holistic approaches in animal ecology. *Oikos* 53, 279–280.

Likens, G.E. (ed.) (1988) *Long-term Studies in Ecology: Approaches and Alternatives*. Springer Verlag, New York.

O'Connor, T. (1999) Community change in rangelands: towards improving our understanding. In: Eldridge, D. and Freudenberger, D. (eds) *People of the Rangelands. Building the Future. Proceedings of the VI International Rangeland Congress*. VI International Rangeland Congress Inc., Townsville, Australia, pp. 203–208.

O'Neill, R.V., DeAngelis, D.L., Waide, J.B. and Allen, T.F.H. (1986) *A Hierarchical Concept of Ecosystems*. Princeton University Press, Princeton, New Jersey.

Passioura, J.B. (1979) Accountability, philosophy and plant physiology. *Search* 10, 347–350.

Pastor, J. and Cohen, Y. (1997) Herbivores, the functional diversity of plant species, and the cycling of nutrients in ecosystems. *Theoretical Population Biology* 51, 165–179.

Rietkerk, M. and van de Koppel, J. (1997) Alternate stable states and threshold effects in semi-arid grazing systems. *Oikos* 79, 69–76.

Rietkerk, M., van den Bosch, F. and van de Koppel, J. (1997) Site-specific properties and irreversible vegetation changes in semi-arid grazing systems. *Oikos* 80, 241–252.

Rosswall, T., Woodmansee, R.G. and Risser, P.G. (eds) (1988) *Scales and Global Change: Spatial and Temporal Variability in Biospheric and Geospheric Processes*. John Wiley & Sons, Chichester, UK.

Schimel, D.S., Emanuel, W., Rizzo, B., Smith, T., Woodward, F.I., Fisher, H., Kittel, T.G.F., McKeown, R., Painter, T., Rosenbloom, N., Ojima, D.S., Parton, W.J., Kicklighter, D.W., McGuire, A.D., Melillo, J.M., Pan, Y., Haxeltine, A., Prentice, C., Sitch, S., Hibbard, K., Nemani, R., Pierce, L., Running, S., Borchers, J., Chaney, J. *et al.* (1997) Continental scale variability in ecosystem processes – models, data, and the role of disturbance. *Ecological Monographs* 67, 251–271.

Schlesinger, W.H. and Pilmanis, A.M. (1998) Plant–soil interactions in deserts. *Biogeochemistry* 42, 169–187.

Scholes, R.J. and Archer, S.R. (1997) Tree–grass interactions in savannas. *Annual Review of Ecology and Systematics* 28, 517–544.

Seastedt, T.R. (1995) Soil systems and nutrient cycles on the North American prairie. In: Joern, A. and Keeler, K.H. (eds) *The Changing Prairie*. Oxford University Press, Oxford, pp. 157–174.

Shugart, H.H., West, D.C. and Emanuel, W.R. (1981) Patterns and dynamics of forests: an application of simulation models. In: West, D.C., Shugart, H.H. and Botkin, D.B. (eds) *Forest Succession: Concepts and Applications*. Springer-Verlag, Heidelberg, pp. 74–94.

Stokes, C.J. (1999) Woody plant dynamics in a south Texas savanna: pattern and process. PhD Dissertation. Texas A&M University, College Station, Texas.

Stroh, J.C., Archer, S., Doolittle, J.A. and Wilding, L.P. (2001) Detection of edaphic discontinuities with ground-penetrating radar and electromagnetic induction. *Landscape Ecology* 16, 377–390.

Tansley, A.G. (1935) The use and abuse of vegetational concepts and terms. *Ecology* 16, 284–307.

Tilman, D. (1989) Ecological experimentation: strengths and conceptual problems. In: Likens, G.E. (ed.) *Long-term Studies in Ecology*. Springer, New York, pp. 136–157.

Tilman, D. and Downing, J.A. (1994) Biodiversity and stability in grasslands. *Nature* 367, 363–365.

Tilman, D. and Wedin, D. (1991) Plant traits and resource reduction for five grasses growing on a nitrogen gradient. *Ecology* 72, 685–700.

Tilman, D., Wedin, D. and Knops, J. (1996) Productivity and sustainability influenced by biodiversity in grassland ecosystems. *Nature* 379, 718–720.

Turner, M.G. and Dale, V.H. (1998) Comparing large, infrequent disturbances: what have we learned? *Ecosytems* 1, 493–496.

Vitousek, P.M. (1982) Nutrient cycling and nutrient use efficiency. *American Naturalist* 119, 553–572.

Vitousek, P.M., Aber, J., Howarth, R.W., Likens, G.E., Matson, P.A., Schindler, D.W., Schlesinger, W.H. and Tilman, G.D. (1997) Human alteration of the global nitrogen cycle: causes and consequences. *Ecological Society of America Issues in Ecology* 1, 15.

Walker, J., Sharpe, P.J.H., Penridge, L.K. and Wu, H. (1989) Ecological field theory: the concept and field tests. *Vegetatio* 83, 81–95.

Ward, D. (1993) Foraging theory, like any other field of science, needs multiple working hypotheses. *Oikos* 67, 376–378.

Watson, I.W., Burnside, D.G. and Holm, A.M. (1996) Event-driven or continuous: which is the better model for managers? *The Rangelands Journal* 18, 351–369.

Wedin, D.A. (1995) Species, nitrogen, and grassland dynamics: the constraints of stuff. In: Jones, C.G. and Lawton, J.H. (eds) *Linking Species and Ecosystems*. Chapman and Hall, New York, pp. 253–262.

Wedin, D.A. (1999) Nitrogen availability, plant–soil feedbacks and grassland stability. In: Eldridge, D. and Freudenberger, D. (eds) *People of the Rangelands. Building the Future. Proceedings of the VI International Rangeland Congress*. VI International Rangeland Congress Inc., Townsville, Australia, pp. 193–197.

Werner, E.E. (1998) Ecological experiments and a research program in community ecology. In: Resetarits, W.J., Jr and Bernardo, J. (eds) *Experimental Ecology: Issues and Perspectives*. Oxford University Press, New York, pp. 3–26.

Wessman, C.A. (1992) Spatial scales and global change: bridging the gap from plots to GCM grid scales. *Annual Review of Ecology and Systematics* 23, 175–200.

Wessman, C.A., Bateson, C.A. and Benning, T.L. (1997) Detecting fire and grazing patterns in tallgrass prairie using spectral mixture analysis. *Ecological Applications* 7, 493–511.

Wilson, E.O. (1998) *Consilience: the Unity of Knowledge*. Knopf, New York.

Yodzis, P. (1988) The indeterminancy of ecological interactions as perceived through perturbation experiments. *Ecology* 69, 508–515.

Range Management and Plant Functional Types

7

Sandra Díaz, David D. Briske and Sue McIntyre

Plant Functional Types: New Developments for an Old Idea

A changing context

One of the recurrent themes in plant ecology and range management is the need to scale information from specific case studies towards broader ecological patterns and processes. Information must be interpreted and applied at larger scales because the most serious challenges to natural resource management operate at regional and global scales. *A central problem encountered when scaling vegetation responses to regional levels is our limited ability to quantify and interpret complex floristic responses involving a large number of individual species.* This provides a strong justification for the development of a more generalized pattern of vegetation responses involving a manageable number of plant groups that have similar life history strategies and responses to environmental stress and disturbance (McIntyre, 1999).

The concept of plant functional types provides a promising tool to bridge the gap between specific, detailed studies and broader scale problems. Plant functional types are sets of plants exhibiting similar responses to environmental conditions and having similar effects on the dominant ecosystem processes (Gitay and Noble, 1997). The classification of plant species into similar groups based on their morphological and physiological traits provides new insights into the dynamics of vegetation change and associated ecological processes.

©CAB *International* 2002. *Global Rangelands: Progress and Prospects* (eds A.C. Grice and K.C. Hodgkinson)

Plant functional groupings are potentially useful communication tools for land managers, who may not necessarily relate to taxonomic units, particularly when dealing with species-rich natural rangelands. As a contribution to the above problem we provide: (i) a description of the concept of plant functional types and their potential contribution to natural resource management; (ii) insight into the ecological basis of plant functional types and ways in which they might be identified; and (iii) an overview of the application and relevance of plant functional types to rangeland management with emphasis on vegetation response in grazed systems.

The concept of plant functional types

The search for a plant classification, which can account for how plants respond to their environment, is at least as old as the Linnaean classification. This is an acknowledgement that plants originating from phylogenetically distant groups may possess similar ecological traits, if they evolve in similar types of environments. The following examples illustrate the importance of environmental selection pressures on plant form and function.

- Succulent, spiny species within a range of families including Cactaceae, Chenopodiaceae, Euphorbiaceae, Fouquieriaceae and Asphodelaceae are well adapted to arid environments.
- Numerous shrub species from unrelated families in the fynbos biome have evolved a low stature, with small, evergreen, sclerophyll leaves (Bond, 1997), suggesting that these traits are strongly selected in this environment.

Conversely, plants with close phylogenetic relationships, growing under very different environmental contexts, can, over evolutionary time, exhibit very different morphology and physiology. For example:

- South American species of *Acacia* are deciduous, with tender, highly palatable compound leaves, while Australian species in the same genus are evergreen, often with tough, mostly unpalatable phyllodes.
- *Salix humboldtiana* is a fast-growing temperate–subtropical tree while *Salix arctica* is a forb-like dwarf that grows in the Arctic tundra.

These examples also provide justification for classifying unrelated taxa on the basis of morphological and physiological traits. *The clear involvement of environmental selection pressures on plant form and function provides the basis for development of plant functional types.* The definition of plant functional types presented here reflects two important issues. First, how do plants respond to selection pressures such as those arising from climate and disturbance? Although similar traits can be selected for in phylogenetically unrelated species, there can be more than one solution or strategy for survival in specific environments. This implies that a number of plant functional types may occur in any particular vegetation type. For example, in the fynbos described above, an alternative plant functional type that co-occurs with the dominant low growing shrubs are grass-like forms, converging in appearance, from the Poaceae, Cyperaceae and Restionaceae (Bond, 1997). However, while such clear examples can exist, the task of identifying plant functional types can be extremely difficult and often involves identifying a small number of discrete 'types' from an array of more or less continuous variation between plant species. A second important issue in the search for plant functional types concerns how plants influence major community and ecosystem processes and, therefore, the main ecosystem services to human populations.

Response Plant Functional Types and Effect Plant Functional Types

Groups of plant species which respond to the abiotic and biotic environment in similar ways can be defined as response plant functional types (Landsberg, 1999; Walker *et al.*, 1999). This approach to the classification of plant functional types is the most common and the oldest one. It attempts to answer the questions: 'What kinds of plants tend to thrive in particular environments?' and, 'How many plant strategies for survival exist in a particular environment?' The most widely recognized and implemented response plant functional type system is the Raunkiaer's Life Form Classification (Raunkiaer, 1934), which groups plants according to the position of the meristems that enable plants to persist through unfavourable seasons. Other examples of widely used response plant functional type classifications are the distinction between 'increaser' and 'decreaser' species based on their response to grazing (Dyksterhuis, 1949), and the prediction of species responses to fire and grazing based on vital plant attributes (Noble and Slatyer, 1980).

On the other hand, *groups of plants which have similar effects on the dominant ecosystem processes, such as productivity, nutrient cycling and trophic transfer, can be defined as effect plant functional types* (Landsberg, 1999; Walker *et al.*, 1999). Effect plant functional types and response plant functional types can, and often do, overlap to various degrees. They can be identified using the same methodological steps (both a priori and a posteriori approaches, see below). The basic distinction between effect plant functional types and response plant functional types is the kind of question underlying the study. In the case of effect plant functional types, the crucial questions are 'What is the functional role of species in ecosystems?' (Lawton, 1994) and 'Does a loss of biodiversity compromise ecosystem function?' (Schulze and Mooney, 1994). These questions address the effect of various plant types on ecosystem dynamics. Circumstantial evidence is accumulating that supports the occurrence of biotic control of ecosystem and landscape processes (Schulze and Mooney, 1994; Mooney et al., 1996; Chapin et al., 1997), but this concept remains the focus of considerable debate. Examples of classification systems based on effect plant functional types are much less common than those based on response plant functional types, although the work of Grime (1977) and Leps et al. (1982) are early examples of the incorporation of ecosystem effects into functional classifications of plants. Subsequent reference to plant functional types will cover both response plant functional types and effect plant functional types unless specified otherwise.

Approaches for Identifying Plant Functional Types

A priori and a posteriori approaches

The application of the plant functional type approach to range management and vegetation assessment obviously requires identification of meaningful plant functional types. Two main approaches exist for their development: (i) define plant functional types before the study is carried out (a priori approach); and (ii) identify plant functional types based on the selection of multiple plant traits at the end of the study (multivariate or a posteriori approach).

Most plant functional type classifications have been determined a priori by selecting a single criterion to define types prior to data collection (e.g. bud position, C_3 or C_4 photosynthetic pathway, grasses or forbs, geophytes, therophytes, or hemicryptophytes). In contrast, a posteriori approaches have become increasingly common in recent years. This approach is based on the identification of multiple traits of numerous species and the important plant traits for developing plant functional types are defined a posteriori, following analysis of these multiple traits. The strength of the a posteriori approach is that it attempts to establish, through statistical correlation, actual links between a putative trait and its functional role in the ecosystem. This approach provides a means for rigorously testing functional classifications.

Several research groups in different parts of the world have taken the a posteriori approach, building extensive databases and defining plant functional types and 'best' traits (e.g. Montalvo et al., 1991; Díaz *et al.*, 1992; Leishman and Westoby, 1992; Chapin *et al.*, 1996; Díaz and Cabido, 1997; Grime *et al.*, 1997; Reich *et al.*, 1997; Wardle *et al.*, 1998; Lavorel *et al.*, 1999). These studies vary in scale from very detailed (e.g. Díaz *et al.*, 1992; Boutin and Keddy, 1993; Golluscio and Sala, 1993) to very coarse (e.g. Chapin *et al.*, 1996; Díaz and Cabido, 1997), and from responses mostly to climate (e.g. Chapin *et al.*, 1996; Díaz and Cabido, 1997; Reich *et al.*, 1997), to resource availability *in situ* (e.g. Golluscio and Sala, 1993; Grime *et al.*, 1997), or to disturbance (Montalvo *et al.*, 1991; McIntyre *et al.*, 1995; Lavorel *et al.*, 1999).

Typically, in the a posteriori approach, numerous traits are initially considered but only a few of them prove useful in defining the main trends of variation among species. Those traits prove to be good candidates for further investigation. In general, vegetative traits (e.g. canopy height, specific leaf area, leaf and plant longevity, position of dormant buds) tend to define clearer groups than regeneration traits (e.g. seed size, dispersal mode, pollination mode, flowering phenology), and the two groups of traits are not consistently related. *It is frequently observed that fundamental plant traits do not vary independently, but rather they tend to be associated in consistent patterns of specialization, or plant syndromes* (see Landsberg *et al.*, 1999, for an exception). This suggests that there are physiological trade-offs between major processes including growth rate, herbivore defence and resource storage that plants are unable to overcome (e.g. Grime, 1977; Chapin, 1980; Coley, 1983). Plants are also assumed to make trade-offs between abundant resources to effectively acquire scarce resources to promote growth and survival (Chapin, 1980). The existence of recurrent sets of highly correlated traits has some practical implications. If these associations among traits can be proven consistent, it would not be necessary to measure all traits in order to identify plant syndromes and develop plant functional types. However, considerable work is still required to be able to confidently predict trait correlations (syndromes) in an unstudied vegetation type.

Selection of important plant traits

The initial selection of plant traits to be considered in a functional type analysis represents a critical step in the search for functional types. Evaluation of the maximum number of variables has not proven to be a good approach because it often results in extensive species lists that make this task operationally unfeasible and/or economically prohibitive for most research groups. The introduction of numerous variables that are strongly correlated (e.g. individual plant biomass and plant height) may lead to severe distortion in the multivariate analysis. Therefore, *a relatively small number of traits is required for a plant functional type methodology that will be adopted by a large number of researchers and that will be sufficiently robust for global comparisons. These traits must meet the following criteria: (i) ecologically relevant with respect to the processes and scale of interest (e.g. climate, soil quality, grazing, fire); (ii) feasibility for rapid and standardized measurement in various regions and vegetation types; and (iii) the procedures must be cost-effective.*

Ideally, the list of plant traits to be evaluated to identify plant functional types would be attained by a general consensus among research groups. *A standardized approach would facilitate global comparisons and increase the potential for identification of general patterns of vegetation responses.* Complete consensus regarding key traits and specific protocols to measure them is unlikely, nor is it totally desirable, as continued exploration of new traits will build our knowledge. Nevertheless, there have been several new contributions to the issue of trait selection, notably Westoby's (1998) leaf–height–seed (LHS) scheme and Hodgson *et al.*'s (1999) operational definition of Grime's CSR scheme. A further contribution to the issue of trait selection was achieved in 1998, as part of GCTE Task 2.2.1, when a diverse group of scientists agreed on a core list of traits considered to be of general importance in the identification of plant functional types (Weiher *et al.*, 1999; Table 7.1). It is important to stress, however, that even a core list of traits will require further refinement in order to usefully describe specific ecological variables (e.g. see grazing-related core traits, McIntyre *et al.*, 1999) or environments. For example, plant phenology can show considerable inter-annual variation in arid

Table 7.1. The common core traits proposed by Weiher *et al.* (1999).

Trait	Function
Seed weight	Dispersal distance, longevity in seed bank, establishment success, fecundity
Seed shape	Longevity in seed bank
Method of seed dispersal	Dispersal distance, longevity in seed bank
Vegetative growth	Space acquisition
Specific leaf area, leaf water content	RGR, plasticity, stress tolerance, evergreenness, leaf longevity
Height	Competitive ability
Above-ground biomass	Competitive ability, fecundity
Life history	Plant longevity, space-holding ability, disturbance tolerance
Onset of flowering	Stress avoidance, disturbance avoidance
Ability to resprout after a disturbance	Disturbance tolerance
Density of wood	Plant longevity, carbon storage

environments that limits its diagnostic value in these regions and C_3 versus C_4 metabolism may not be a useful trait in temperate-cool regions, but it may be a key trait for modelling large-scale data sets.

Calibration between 'soft' and 'hard' traits

As discussed above, there is a strong case for the adoption of easily measured structural–functional traits, such as those proposed by Box (1996) or Díaz and Cabido (1997). This is a legitimate approach in its own right. However, its strength would be substantially increased if the 'soft' (i.e. easily measured) traits could be calibrated against 'hard' traits that have a direct and well-established relationship to ecosystem function (e.g. decomposition rate, nutrient content) but evaluating them is often time-consuming. The process of calibration between 'soft' and 'hard' traits is in the early stages of development in most regions of the world, although some promising results have recently emerged. For instance, Díaz *et al.* (1999b) showed a significant association between the 'soft' traits of specific leaf area and leaf toughness and the 'hard' traits of leaf N content, palatability for invertebrate herbivores and decomposition rate, over a very wide spectrum of plant families and growth forms. This process of calibration among traits may substantially enhance the applicability of plant functional types to various management applications by making the process more cost-effective.

Plant Functional Response Types: Climate and Disturbance History

Although the concept of response plant functional types is intuitively appealing, and they have been sought since the earliest days of ecology, it is still difficult to predict what response plant functional types will predominate under different frequencies and/or intensities of disturbance, such as grazing, fire or flooding. For example, the presence of aerenchyma in roots and of a persistent seed bank seem important traits in defining response plant functional types in areas with seasonal droughts and floods; the presence of thick bark, lignotubers and serotinous seeds seem important in fire-prone areas; and leaf toughness and nutrient content, architectural plasticity and bud position seem important in some areas chronically subjected to grazing. As documented by Noy-Meir and Sternberg (1999), even two disturbances which involve removal of above-ground biomass, such as grazing and fire, are associated with different sets of plant traits in the same geographical area.

An added level of complexity in trying to identify key plant traits associated with disturbance response is the consideration of disturbance history. The predominant plant groups or plant traits which appear under different grazing intensities, for example, seem to depend strongly on a combination of climate and evolutionary history of herbivory (Milchunas et al., 1988; Milchunas and Lauenroth, 1993; Díaz et al., 1999a). Different plant traits can predominate in areas with the same annual precipitation and livestock density, depending on whether grazing has been a strong selective pressure over evolutionary time or not. For example, heavy grazing is associated with high abundance of annual species in many regions. In the Eastern Mediterranean, however, heavy grazing promotes geophytes and legumes (Hadar et al., 1999). Annuals do not substantially increase, since they represent most of the biomass in both grazed and ungrazed sites in this region, which has a very long evolutionary history of grazing (Perevolotsky and Seligman, 1998).

In summary, the question of 'What plant traits or types predominate under different grazing regimes?' does not have a straightforward answer. Most of the published evidence is related to very specific cases, is at the species level, and is scattered in the ecological, agronomic and phytosociological literature. At present, and before a more comprehensive framework is developed, the traits and response plant functional types likely to predominate in the face of different land-management situations are likely to be strongly site-specific, as illustrated in the next section.

Plant Functional Response Types in Grazed Systems

The application of response plant functional types to describe vegetation responses to grazing for application in range management and grassland agriculture is not a novel concept. The most recognized and most widely applied response plant functional type for grazing application is the concept of increaser and decreaser species associated with range condition and trend analysis (Dyksterhuis, 1949). A closely associated response plant functional type classification is based on the distinction among grass growth forms including short, mid and tall grasses (Arnold, 1955). These response plant functional types were initially based on empirical data, but specific plant traits were subsequently associated with species responses to grazing (Hendon and Briske, 1997; Briske, 1999). In fact, it is often assumed that grazing resistance is based on the occurrence of a relatively small number of traits, or even a single trait, associated with the developmental morphology or physiological function of individual species (Simms, 1992).

These response plant functional types have provided a large amount of valuable information concerning the relative responsiveness of plants to grazing. Several plant traits, including the location and availability of meristems, architectural attributes influencing palatability and residual leaf area following defoliation, have proved especially important to our understanding of grazing resistance in plants (Briske and Richards, 1995). However, an increasing number of cases exist where grazing resistance has not been effectively explained by the presence or absence of specific plant traits (e.g. Hendon and Briske, 1997). Although both traditional response plant functional type systems mentioned above are still used to various degrees, it has become clear that the plant traits traditionally associated with grazing resistance are not always sufficient to predict or interpret species responses to grazing (Noy-Meir and Sternberg, 1999). The limitations encountered by these traditional response plant functional types can be organized into three general categories: (i) the existence of multiple categories of resistance traits; (ii) the occurrence of trade-offs among categories of resistance traits; and (iii) the disproportionate expression of categories of resistance traits at various ecological scales (Briske, 1999).

Multiple categories of resistance

It has previously been emphasized that plant adaptation to stress and disturbance often involves the evolution of multiple traits (Grime, 1977). Similarly, recognition of several strategies of resistance to grazing demonstrates that more than a single trait is involved in determining the grazing resistance of plants (Simms, 1992; Briske, 1996). Inordinate emphasis on a small number of specific plant traits may have inadvertently diverted attention from the identification and interpretation of more pervasive strategies of grazing resistance.

Temporal variation displayed by various resistance traits further challenges the development of effective response plant functional types to evaluate plant responses to grazing. The dynamic expression of morphological and physiological traits, including canopy architecture, various inducible defences and compensatory physiological processes, has been documented within and between species (Briske and Richards, 1995; Briske, 1996). Species with a high degree of phenotypic plasticity may even shift the expression of grazing resistance from a tolerance to an avoidance strategy with an increasing intensity of grazing (Hodgkinson *et al.*, 1989). In other cases, traits may be difficult to recognize because they represent more of a life history expression than a distinct morphological or physiological trait. This is clearly illustrated by the 'phenological trait' associated with the extended display of green biomass by perennial grasses to annual grasses in Mediterranean grasslands (Noy-Meir and Sternberg, 1999). Examples of the dynamic expression of various resistance traits indicate that grazing resistance does not represent a static value that can invariably be assigned to individual species in all situations.

Effective interspecific comparisons of grazing resistance are constrained by our inability to incorporate multiple resistance traits into a standardized expression applicable to various growth forms, life history strategies and phenological stages. Grazing tolerance is by definition based on the rate or magnitude of biomass production following defoliation (Rosenthal and Kotanen, 1994; Briske, 1996). However, insight necessary to prioritize or weight the various traits and processes associated with growth following defoliation, including leaf and shoot number, canopy height and volume, biomass partitioning to various organs and reproductive effort, is very limited. *A limited understanding of the relative importance of specific plant traits has impeded the development of a comprehensive interpretation of grazing resistance in plants.*

Trade-offs among resistance categories

The expression of grazing resistance by plants can be divided into tolerance and avoidance strategies (Rosenthal and Kotanen, 1994; Briske, 1996). The tolerance strategy promotes rapid leaf replacement following defoliation while the avoidance strategy minimizes the frequency and intensity of grazing. Although both strategies contribute to grazing resistance, the specific traits involved are very likely unique to a particular strategy (Westoby, 1999). Tolerance is associated with traits that contribute to rapid leaf replacement following defoliation while avoidance is associated with traits that defend plants from grazing. An attempt to identify traits without recognition of the unique strategy involved would very likely produce inconsistent results as the absolute expression of tolerance and avoidance varies among species (Westoby, 1999). For example, decreaser species may rely on tolerance mechanisms for grazing resistance to a greater extent than increaser species because tolerance traits are closely correlated with the competitor strategy of the first group (Briske, 1996). In contrast, large investments in grazing avoidance may divert resources from growth and potentially reduce the expression of grazing tolerance. The expression of greater tolerance by dominant compared to subordinate species can be suppressed prior to the expression of avoidance mechanisms because grazers can potentially remove biomass more rapidly than it can be replaced by tolerance mechanisms. Selective grazing of dominants compared to subordinate species would potentially shift the competitive advantage from dominants to subordinates and induce a shift in species composition.

The tolerance and avoidance strategies of grazing resistance have only recently been placed in a conceptual framework to hypothesize that the relative expression of these two strategies may determine the productivity and composition of grazed plant communities (Augustine and McNaughton, 1998). Grazing tolerance is assumed to be of equal or greater importance than grazing avoidance (i.e. selective grazing) in systems where highly palatable species retain dominance. The reverse is assumed to be the case in systems where dominants are replaced by subordinate species. Although this interpretation is highly plausible, the relative expression of grazing tolerance and avoidance has not been quantified for most species or vegetation types and the potential trade-offs between these two strategies of grazing resistance are only currently being considered (Mauricio et al., 1997).

Expression of resistance at various scales

The initial response plant functional type classifications analysed in the previous section were developed in the mid-20th century, when a more reductionist view of science prevailed. *Emphasis on individual plant traits associated with grazing resistance makes is difficult to incorporate and interpret associated processes occurring at higher ecological scales* (Briske, 1999). There is increasing recognition that *individual plant responses to grazing may not directly scale up to communities or landscapes because grazing conveys indirect, as well as direct effects.* Indirect grazing effects involve both biotic and abiotic processes external to plants in contrast to the direct removal of photosynthetic and meristematic tissues (McNaughton, 1983). Important indirect effects known to mediate plant responses to grazing, include selective grazing among species (Anderson and Briske, 1995), grazing-modified competitive interactions (Caldwell et al., 1987) and drought-grazing interactions (O'Connor, 1994). Indirect grazing effects may be of equal or greater importance than the direct effects of grazing in determining vegetation responses, but they are often minimized or excluded from investigations designed to assess grazing resistance based on specific plant traits.

Specific issues of temporal and spatial scale introduce additional complexity in the process of developing response plant functional types for grazing resistance. For example, grazing tolerance is often assessed by evaluating the short-term regrowth responses of plants while traits contributing to plant persistence over the long term often receive less attention. For example, rapid leaf replacement by new leaf initiation will promote tolerance, but if tiller initiation is suppressed following grazing, long-term plant persistence may be compromised. The spatial association among plants expressing various degrees of grazing avoidance may also influence the frequency and intensity of grazing at the patch scale. The protection a palatable plant derives from close association with unpalatable plants is referred to as associative defence (Hay, 1986). This may partially explain why a plant species may respond negatively to grazing in one environment, but positively in another environment. The relative expressions of grazing avoidance among various plant species may affect patterns of selective grazing without necessitating a change in the absolute expression of avoidance by individual species.

The challenges associated with the identification of response plant functional types for determining vegetation responses to grazing are great and further substantiate several cautionary considerations made previously. Response plant functional types developed for grazing will have to be regionally specific and clearly specify the ecological and managerial information to be provided and the scale at which it is to be applied. It is anticipated that the development of effect plant functional types will follow development of response plant functional types and will more effectively incorporate the indirect effects of grazing within ecosystems. *The challenges involved in selecting and interpreting plant traits associated with grazing responses must be recognized and addressed prior to the development of more effective plant functional types for grazing applications.* A synthesis of published work, considering different climates, vegetation types and evolutionary histories of grazing may represent an important step towards this goal.

Plant Functional Effect Types and Ecosystem Function

In contrast to the situation of response plant functional types and different combinations of climate, history and land use, *effect plant functional types and major ecosystem processes seem to be much more consistently linked across different ecosystem types.* Local dominance by species with specific plant traits appears to directly influence various ecosystem processes including productivity, nutrient cycling, trophic transfer, temperature buffering, flammability, etc. (Table 7.2). Sufficient evidence exists to support an evaluation, at least in comparative terms, of the magnitude, direction and rate of some ecosystem processes on the basis of the traits associated with the dominant species (see Díaz et al., 1999c, for further evidence and discussion). It is important to stress, however, that links between plant traits and ecosystem function are much better understood in the case of vegetative traits (e.g. fast-growing plants with nutrient-rich leaves are associated

Table 7.2. Examples of individual plant traits that may influence processes of the community/ecosystem in which they are dominant. See Díaz *et al.* (1999c) for references and more detailed explanation.

Individual traits	Community/ecosystem processes
Relative growth rate	Productivity
Leaf turnover rate	Nutrient cycling
	Production efficiency
Nutrient content	Nutrient cycling
	Carrying capacity for herbivores
Biomass	Flammability
Life span	Resistance
Canopy structure	Water interception and runoff
	Temperature buffering
	Soil stability
Secondary growth	Carbon sequestration
Root architecture	Water uptake
Reserve organs	Resilience
Pollination mode	Expansion over landscape
Persistent seed bank	Resilience
Seed number	Expansion over landscape
Dispersal mode	Expansion over landscape
Presence of root symbionts	Nutrient cycling
(e.g. mycorrhizae)	Rate of succession

with high decomposition rate and high productivity at the ecosystem level) than in the case of regenerative traits (such as phenology or pollination and seed dispersal modes).

It has been recognized that the amount of plant biomass, regardless of species composition, has a strong influence on ecosystem function. However, it appears that species richness confers resilience to plant communities (Landsberg, 1999; Walker *et al.*, 1999). Therefore, effect plant functional types defined on the basis of biomass production may contain a wide range of species that are capable of maintaining similar productivity. This is based on the concept of functional redundancy (Walker *et al.*, 1999) which implies that members of the same effect plant functional type can perform similar ecosystem functions. Consequently, *species may replace each other to varying degrees without a loss of ecosystem function because functionally equivalent species represent a greater range of ecological tolerances to buffer environmental changes.*

Once the main response plant functional types have been identified for a region, it should be possible to translate them into effect plant functional types to analyse the consequences of dominance by specific plant groups for at least the most obvious ecosystem processes and services. These may include the magnitude and seasonality of biomass production, carrying capacity for livestock and wildlife, flammability, water retention and soil protection. The number of species within each effect plant functional type should provide an indication of the resilience of the ecosystem processes as previously indicated. Therefore, the degree of species and trait specificity required in plant functional types is dependent upon the ecological and managerial information sought. If the focus is on present-day ecosystem performance, an evaluation of only the main dominants (e.g. those with > 10% cover) should be sufficient. However, if ecosystem resilience to various disturbance regimes is being evaluated over the long term, species richness within plant functional types would become highly relevant and an evaluation of the entire local flora may be appropriate.

Resolution of Plant Functional Types: Regional vs. Global Classifications

The existence of consistent plant specialization patterns and trade-offs between plant processes does not necessarily mean that we should seek a single plant functional type system that would be appropriate for all applications and scales (Gitay and Noble, 1997; Lavorel et al., 1997; Lavorel and McIntyre, 1999). On the other hand, the utility of developing numerous specific-purpose functional type schemes has not been demonstrated either (Westoby, 1999). Applications concerning plant responses to general climatic conditions or resource availability would likely be best served by general allocation models with a small set of extreme types (e.g. Grime, 1977; Chapin, 1980). This approach may also represent the appropriate resolution required for development of various global vegetation models aimed at predicting vegetation responses to global change at a continental scale. On the other hand, that approach would be too coarse for local and regional management and conservation planning. In these cases, much more detailed plant functional types are required and they must be based on traits specifically tailored to the local environment and land-use considerations. In summary, *a nested hierarchy of plant functional types seems to be the most reasonable answer, since the challenges to be faced are multi-scale.* As recently pointed out by Grime (1998), *a small set of very general plant functional types, based in trade-off models, appears to be more appropriate to the prediction of vegetation processes at the global and trans-regional levels, whereas a much higher number of more precise plant functional types seems most useful for land-use planning at the regional to local levels.*

The Relevance of Plant Functional Types to Rangeland Management

The process of identifying plant functional types appears to be a highly academic exercise. It is reasonable then to ask whether they have any real relevance to rangeland management and managers. In so far as all effective communications require a simplification of complex information, we know that we need to identify plant functional types. We still have to identify useful groups. Questions such as 'Which biological features favour and which features disadvantage plants under grazing?' or, 'What are the main factors that determine the promotion of different plant traits by grazing?' are fundamental for appropriate management and conservation, as well as for the progress of range science, and have never been answered at a global scale. The plant functional type approach is arguably the only way forward to address these kinds of questions.

The relevance of plant functional types to rangeland ecology and management is clearly illustrated by previous attempts to develop these classifications and their continued use given the recognition that their application may not be appropriate in some cases, e.g. increaser/ decreaser species. *The relevance of plant functional types resides in the need to integrate and generalize site-specific information to broader scales for management applications.* Provided that management questions are clearly defined in the context of a specific region, it should be relatively straightforward to identify general response plant functional types using key plant traits. *The plant functional type approach may prove to be most useful in rangeland planning and management at regional scales.* Specifically, it may provide greater insight into issues of vegetation response and their potential impacts on ecosystem function.

The major application of plant functional types at regional scales will be for monitoring ecosystem structure and function to evaluate ecological impacts and determine appropriate management responses. A simple example is the use of perennial tussock grasses as a response plant functional type relating to grazing in tropical grasslands and a effect plant functional type in terms of the capacity of perennial grasses to contribute to soil health and soil and water capture (Tongway and Hindley, 1995). This application is relevant to both land management agencies and producer groups.

Land managers may utilize plant functional types as both indicators of vegetation change associated with management activities and environmental changes as well as indicators of the sustainability of rangeland ecosystems. A potentially powerful aspect of plant functional types is that they do not require that land managers acquire in-depth taxonomic knowledge of the flora in a region, but rather rely on critical information associated with unique plant groups that may be intuitively simple to assess (e.g. stoloniferous, low-growing, early flowering). This information is more likely to be accessible to range managers and communities with a limited technical base. In some situations, local communities already have an intuitive, but sometimes surprisingly precise, knowledge of the biological traits possessed by different plants. An important future development will be to link plant functional types with important ecosystem functions to more effectively monitor rangelands and assess land-use and climate change on rangeland sustainability.

Rangeland monitoring, assessment and policy development on a continental scale will also require simplified models of vegetation change, and major efforts to identify plant functional types have been associated with global-scale modelling (Smith *et al.*, 1997). *Identification of appropriate plant functional types provides an essential underpinning of the knowledge relevant to national and international planning and policy-making.*

Acknowledgements

Universidad Nacional de Córdoba (Res. 263/95–177/97), IAI ISP I and III, CONICET (PIA 7706 and PEI 7706), Fundación Antorchas (A-13532/1–95), CONICOR (347/95), Darwin Initiative (DETR UK) and the USDA-NRI Ecosystem Science Program supported work contributing to this synthesis (grant #95–37101–2029 to DDB). We are very grateful to the organizers of the VI International Rangeland Congress (IRC). The participants in the IRC Symposium on Range Management and Plant Functional Types and in the Workshop on Plant Functional Types and Grazing (Brisbane) actively contributed to the development of concepts and ideas expressed in this chapter. This is a contribution to IGBP-CGTE Task 2.2.1.

References

Anderson, V.J. and Briske, D.D. (1995) Herbivore-induced species replacement in grasslands: is it driven by herbivory tolerance or avoidance? *Ecological Applications* 5, 1014–1024.

Arnold, J.F. (1955) Plant life-form classification and its use in evaluating range conditions and trend. *Journal of Range Management* 8, 176–181.

Augustine, D.J. and McNaughton, S.J. (1998) Ungulate effects on the functional species composition of plant communities: herbivore selectivity and plant tolerance. *Journal of Wildlife Management* 62, 1165–1183.

Bond, W.J. (1997) Functional types for predicting changes in biodiversity: a case study in Cape fynbos. In: Smith, T.M., Shugart, H.H. and Woodward, F.I. (eds) *Plant Functional Types*. Cambridge University Press, Cambridge, pp. 174–194.

Boutin, C. and Keddy, P.A. (1993) A functional classification of wetland plants. *Journal of Vegetation Science* 4, 591–600.

Box, E.O. (1996) Plant functional types and climate at the global scale. *Journal of Vegetation Science* 7, 309–320.

Briske, D.D. (1996) Strategies of plant survival in grazed systems: a functional interpretation. In: Hodgson, J. and Illius, A.W. (eds) *The Ecology and Management of Grazed Systems*. CAB International, Wallingford, UK, pp. 37–67.

Briske, D.D. (1999) Plant traits determining grazing resistance: why have they proved so elusive? In: Eldridge, D. and Freudenberger, D. (eds) *People of the Rangelands. Building the Future. Proceedings of the VI International Rangeland Congress*. VI International Rangeland Congress Inc., Townsville, Australia, pp. 901–905.

Briske, D.D. and Richards, J.H. (1995) Plant responses to defoliation: a physiological, morphological and demographic evaluation. In: Bedunah, D.J. and Sosebee, R.E. (eds) *Wildland Plants: Physiological Ecology and Developmental Morphology*. Society for Range Management, Denver, Colorado, pp. 635–710.

Caldwell, M.M., Richards, J.H., Manwaring, J.H. and Eissenstat, D.M. (1987) Rapid shifts in phosphate acquisition show direct competition between neighbouring plants. *Nature* 327, 615–616.

Chapin, F.S., III (1980) The mineral nutrition of wild plants. *Annual Review of Ecology and Systematics* 11, 233–260.

Chapin, F.S., III, Bret-Harte, M.S., Hobbie, S. and Zhong, H. (1996) Plant functional types as predictors of the transient response of arctic vegetation to global change. *Journal of Vegetation Science* 7, 347–357.

Chapin, F.S., III, Walker, B.H., Hobbs, R.J., Hooper, D.U., Lawton, J.H., Sala, O.E. and Tilman, D. (1997) Biotic control of the functioning of ecosystems. *Science* 277, 500–503.

Coley, P.D. (1983) Herbivory and defensive characteristics of tree species in a lowland tropical forest. *Ecological Monographs* 53, 209–233.

Díaz, S. and Cabido, M. (1997) Plant functional types and ecosystem function in relation to global change. *Journal of Vegetation Science* 8, 463–474.

Díaz, S., Acosta, A. and Cabido, M. (1992) Morphological analysis of herbaceous communities under different grazing regimes. *Journal of Vegetation Science* 3, 689–696.

Díaz, S., Cabido, M., Zak, M., Martínez-Carretero, E. and Araníbar, J. (1999a) Plant functional traits, ecosystem structure, and land-use history along a climatic gradient in central-western Argentina. *Journal of Vegetation Science* 10, 651–660.

Díaz, S., Pérez-Harguindeguy, S., Vendramini, F., Basconcelo, S., Funes, G., Gurvich, D., Cabido, M., Cornelissen, J.H.C. and Falczuk, V. (1999b) Plant traits as links between ecosystem structure and functioning. In: Eldridge, D. and Freudenberger, D. (eds) *People of the Rangelands. Building the Future. Proceedings of the VI International Rangeland Congress*. VI International Rangeland Congress Inc., Townsville, Australia, pp. 896–901.

Díaz, S., Cabido, M. and Casanoves, F. (1999c) Functional implications of trait–environment linkages in plant communities. In: Weiher, E. and Keddy, P.A.

(eds) *The Search for Assembly Rules in Ecological Communities.* Cambridge University Press, Cambridge, pp. 338–362.

Dyksterhuis, E.J. (1949) Condition and management of range land based on quantitative ecology. *Journal of Range Management* 2, 104–115.

Gitay, H. and Noble, I.R. (1997) What are plant functional types and how should we seek them? In: Smith, T.M., Shugart, H.H. and Woodward, F.I. (eds) *Plant Functional Types.* Cambridge University Press, Cambridge, pp. 3–19.

Golluscio, R. and Sala, O.E. (1993) Plant functional types and ecological strategies in Patagonian forbs. *Journal of Vegetation Science* 4, 839–846.

Grime, J.P. (1977) Evidence for the existence of three primary strategies in plants and its relevance to ecological and evolutionary theory. *American Naturalist* 111, 1169.

Grime, J.P. (1998) Plant functional types and ecosystem processes. *Mededelingen van de KNAW* 1998, 104–108.

Grime, J.P., Thompson, K., Hunt, R., Hodgson J.G., Cornelissen, J.H.C., Rorison, I.H., Hendry, G.A.F., Ashenden, T.W., Askew, A.P., Band, S.R., Booth, R.E., Bossard, C.C., Campbell, B.D., Cooper, J.E.L., Davidson, A.W., Gupta, P.L., Hall, W., Hand, D.W., Hannah, M.A., Hillier, S.H., Hodkinson, D.J., Jalili, A., Liu, Z., Mackey, J.M.L., Matthews, N., Mowforth, M.A., Neal, A.M., Reader, R.J., Reiling, K., Ross-Fraser, W., Spencer, R.E., Sutton, F., Tasker, D.E., Thorpe, P.C. and Whitehouse, J. (1997) Integrated screening validates primary axes of specialization in plants. *Oikos* 79, 259–281.

Hadar, L., Noy-Meir, I. and Perevolotsky, A. (1999) The effect of shrub clearing and intensive grazing on the composition of a Mediterranean plant community at the functional group and the species level. *Journal of Vegetation Science* 10, 673–682.

Hay, M.E. (1986) Associational plant defenses and the maintenance of species diversity: turning competitors into accomplices. *American Naturalist* 128, 617–641.

Hendon, B.C. and Briske, D.D. (1997) Demographic evaluation of a herbivory-sensitive perennial bunchgrass: does it possess an Achilles heel? *Oikos* 80, 8–17.

Hodgkinson, K.C., Ludlow, M.M., Mott, J.J. and Baruch, Z. (1989) Comparative responses of the savanna grasses *Cenchrus ciliaris* and *Themeda triandra* to defoliation. *Oecologia* 79, 45–52.

Hodgson, J.G., Wilson, P.J., Hunt, R., Grime, J.P. and Thompson, K. (1999) Allocating C-S-R plant functional types: a soft approach to a hard problem. *Oikos* 85, 282–296.

Landsberg, J. (1999) Response and effect – different reasons for classifying plant functional types under grazing. In: Eldridge, D. and Freudenberger, D. (eds) *People of the Rangelands. Building the Future. Proceedings of the VI International Rangeland Congress.* VI International Rangeland Congress Inc., Townsville, Australia, pp. 911–915.

Landsberg, J., Lavorel, S. and Stol, J. (1999) Grazing response groups among understorey plants in arid rangelands. *Journal of Vegetation Science* 10, 683–696.

Lavorel, S. and McIntyre, S. (1999) Plant functional types: is the real world too complex? In: Eldridge, D. and Freudenberger, D. (eds) *People of the Rangelands.*

Building the Future. Proceedings of the VI International Rangeland Congress. VI International Rangeland Congress Inc., Townsville, Australia, pp. 905–911.

Lavorel, S., McIntyre, S., Landsberg, J. and Forbes, T.D.A. (1997) Plant functional classifications: from general groups to specific groups based on response to disturbance. *Trends in Ecology and Evolution* 12, 474–478.

Lavorel, S., McIntyre, S. and Grigulis, K. (1999) Response to disturbance in a Mediterranean annual grassland: how many functional groups? *Journal of Vegetation Science* 10, 661–672.

Lawton, J.H. (1994) What do species do in ecosystems? *Oikos* 71, 367–374.

Leishman, M.R. and Westoby, M. (1992) Classifying plants into groups on the basis of associations of individual traits – evidence from Australian semi-arid woodlands. *Journal of Ecology* 80, 417–424.

Leps, J., Osbornová-Kosinová, J. and Rejmánek, M. (1982) Community stability, complexity and species life-history strategies. *Vegetatio* 50, 53–63.

Mauricio, R., Rausher, M.D. and Burdick, D.S. (1997) Variation in the defence strategies of plants: are resistance and tolerance mutually exclusive? *Ecology* 78, 1301–1311.

McIntyre, S. (1999) Plant functional types: recent history and current developments. In: Eldridge, D. and Freudenberger, D. (eds) *People of the Rangelands. Building the Future. Proceedings of the VI International Rangeland Congress.* VI International Rangeland Congress Inc., Townsville, Australia, pp. 891–893.

McIntyre, S., Lavorel, S. and Trémont, R. (1995) Plant life-history attributes: their relationship to disturbance response in herbaceous vegetation. *Journal of Ecology* 83, 31–44.

McIntyre, S., Lavorel, S., Landsberg, J. and Forbes, T.D. (1999) Disturbance response in vegetation – gaining a global perspective on functional traits. *Journal of Vegetation Science* 10, 621–630.

McNaughton, S.J. (1983) Compensatory plant growth as a response to herbivory. *Oikos* 40, 329–336.

Milchunas, D.G. and Lauenroth, W.K. (1993) Quantitative effects of grazing on vegetation and soils over a global range of environments. *Ecological Monographs* 63, 327–366.

Milchunas, D.G., Sala, O.E. and Lauenroth, W.K. (1988) A generalized model of the effects of grazing by large herbivores on grassland community structure. *American Naturalist* 132, 87–106.

Montalvo, J., Casado, M.A., Levassor, C. and Pineda, F.D. (1991) Adaptation of ecological systems: compositional patterns of species and morphological and functional traits. *Journal of Vegetation Science* 2, 655–666.

Mooney, H.A., Cushman, J.H., Medina, E., Sala, O.E. and Schulze, E.-D. (1996) *Functional Roles of Biodiversity: a Global Perspective.* John Wiley & Sons, New York.

Noble, I.R. and Slatyer, R.O. (1980) The use of vital attributes to predict successional changes in plant communities subject to recurrent disturbances. *Vegetatio* 43, 5–21.

Noy-Meir, I. and Sternberg, M. (1999) Grazing and fire response, and plant functional types in Mediterranean grasslands. In: Eldridge, D. and Freudenberger, D. (eds) *People of the Rangelands. Building the Future. Proceedings of the VI International Rangeland Congress.* VI International Rangeland Congress Inc., Townsville, Australia, pp. 916–921.

O'Connor, T.G. (1994) Composition and population responses of an African savanna grassland to rainfall and grazing. *Journal of Applied Ecology* 31, 155–171.

Perevolotsky, A. and Seligman, N.G. (1998) Role of grazing in Mediterranean rangeland ecosystems. *Bioscience* 48, 1007–1017.

Raunkiaer, C. (1934) *The Life Forms of Plants and Statistical Plant Geography; Being the Collected Papers of C. Raunkiaer.* Translated into English by Carter, H.G., Tansley, A.G. & Miss Fansboll. Clarendon Press, Oxford.

Reich, P.B., Alters, M.B. and Elsworth, D.S. (1997) From tropics to tundra: global convergence in plant functioning. *Proceedings of the National Academy of Sciences USA* 94, 13730–13734.

Rosenthal, J.P. and Kotanen, P.M. (1994) Terrestrial plant tolerance to herbivory. *Trends in Ecology and Evolution* 9, 145–148.

Schulze, E.-D. and Mooney, H.A. (1994) *Biodiversity and Ecosystem Function.* Springer, Berlin.

Simms, E.L. (1992) Costs of plant resistance to herbivory. In: Fritz, R.S. and Simms, E.L. (eds) *Plant Resistance to Herbivory and Pathogens.* University of Chicago Press, Chicago, pp. 392–425.

Smith, T.M., Shugart, H.H. and Woodward, F.I. (1997) *Plant Functional Types.* Cambridge University Press, Cambridge.

Tongway, D. and Hindley, N. (1995) *Manual for Soil Condition Assessment of Tropical Grasslands.* CSIRO Wildlife and Ecology, Canberra.

Walker, B.H., Kinzig, A. and Landsberg, J.J. (1999) Ecosystem function and plant attribute diversity: the nature and significance of dominant and minor species. *Ecosystems* 2, 95–113.

Wardle, D.A., Barker, G.M., Bonner, K.I. and Nicholson, K.S. (1998) Can comparative approaches based on plant ecophysiological traits predict the nature of biotic interactions and individual plant species effects in ecosystems? *Journal of Ecology* 86, 405–420.

Weiher, E., van der Werf, A., Thompson, K., Roderick, M., Garnier, E. and Eriksson, O. (1999) Challenging Theophrastus: a common core list of plant traits for functional ecology. *Journal of Vegetation Science* 10, 609–620.

Westoby, M. (1998) A leaf–height–seed (LHS) plant ecology strategy scheme. *Plant and Soil* 199, 213–227.

Westoby, M. (1999) The LHS strategy in relation to grazing and fire. In: Eldridge, D. and Freudenberger, D. (eds) *People of the Rangelands. Building the Future. Proceedings of the VI International Rangeland Congress.* VI International Rangeland Congress Inc., Townsville, Australia, pp. 893–896.

People and Plant Invasions of the Rangelands

8

Mark Lonsdale and Sue Milton

Introduction

Invasions by exotic plants cause production losses amounting to at least A\$3 billion year^{-1} to the Australian economy, and US\$13 billion to the US economy. It is not widely realized, however, that the majority of weeds probably result from deliberate plant introductions (e.g. Lonsdale, 1994). In addition, human activities – through cultivation, dispersal, disturbance, nitrification and habitat fragmentation – influence the rate at which species become naturalized, and naturalized species become weeds. There is a perception that weeds are a natural disaster like cyclones, whereas in truth they are more a product of our own 'primal urges'. Just as the problem is one of our making, however, the solutions also lie in human hands, and we conclude this chapter by surveying recent developments in weed control.

Why are Rangeland Weeds Important?

Many of the world's rangelands are infested with weeds. *Typical rangeland weeds are shrubs and grasses of low palatability.* Though it is difficult to calculate their economic impact or even the extent of the infestations, losses of production in agricultural systems due to weeds are believed to be in the order of 10% (Combellack, 1989), and there is no reason to suppose that rangelands, which are less intensively managed, would be any less impacted. Farmers and ranchers in the US spend about US$5 billion per annum on weed control, and weed losses to crop and rangeland productivity in the US exceed US$7 billion per annum (Babbitt, 1998).

The area of weed-infested rangeland is increasing. This is because there are few economically viable solutions to plant invasions of the rangelands, and the invading species tend to be long-lived perennials that form self-perpetuating stands able to dominate the plant community and resist further colonization by other species. *Occasionally we make inroads into rangeland weeds through less cost-intensive technologies such as fire and biocontrol, but globally, the weeds are winning the war.*

Although weeds have impacted on people all over the rangelands, it is perhaps in South Africa that the effects on humanity have not only been the most acute and most clearly documented, but are also combatted by people power in the most dramatic fashion (Van Wilgen and van Wyk, 1999). About 750 tree and 8000 other plant species have been introduced to South Africa. At least 161 species are regarded as invasive, and they impact on 10 million ha (8%) of the country. Fuel loads at invaded sites are increased tenfold, increasing fire intensities and causing soil damage, increased erosion and decreased germination from indigenous seed pools. One particularly crucial effect of exotic plants has been that the tall exotic vegetation takes up more groundwater than the short native shrubland, reducing the flow of water in river systems. Other economic impacts have not been well studied, but, for the South African government, the impact on water availability alone justified intervention. Studies in the Western Cape showed that clearing alien plants from catchment areas can deliver additional water at only 14% of the cost of building a new dam. As a consequence, the government set up the 'Working for Water' programme. Over 200 projects across South Africa engage unemployed people in the manual and chemical clearance of exotic stands. Biological control also plays a part in killing stands of some species, which can then be burnt or cleared, and in preventing spread of other species by attacking their seeds.

The justification for this programme is primarily economic, but the same exotic species affect biodiversity. Exotic plants could eliminate several thousand species of plants if spread is not controlled, seriously affecting the delivery of ecosystem services. South Africa has been fortunate in being able to raise significant funding for control programmes through a combination of economic argument and strong political support. However, intersectoral conflicts with the forest industry and other users of plant products still must be resolved. This will require political intervention.

Why do we Import Species?

Plant invasions have for centuries been one of the consequences of human migration and settlement. Any attempt to understand, predict and potentially modify invasions in the future requires that we first recognize the role of human behaviour (Mack, 1999).

Settlers to new lands have always had only two basic options for satisfying their demands for food, fibre, animal forage and any other real or perceived needs from nature. They could either derive these commodities from the native biota or transport the germplasm and cultivation technology to establish species from their homeland in the new locale. Mack (1999) contends that the extent to which they pursue the second option depends on basic human needs, decision-making and a sense of well-being.

To satisfy urgent needs, European colonists frequently sought introductions from homelands rather than conduct a thorough search of the native biota. As colonists in a foreign land where emergency relief was either unreliable or non-existent, they needed to become self-sufficient quickly. Self-sufficiency meant the rapid establishment of a reliable food base that was only minimally dependent on hunting and gathering. For European settlers, crops and domesticated animals that served so well in Europe were a logical solution and were repeatedly requested.

Much less understandable is the continuation of this mindset into modern times, long after the more familiar species have been introduced. Governmental agencies in the late 20th century have continued to practise a liberal policy of plant introduction, despite ample evidence of potentially adverse consequences. In addition, *the numbers of imported species have been swollen, not by agricultural species, but by frivolous ornamental plant introductions* (Panetta, 1993). The market today for ornamentals is ephemeral and unpredictable, although the financial returns for being the first and sole source for 'trend-setting' new introductions are lucrative. Such a market prompts some commercial and amateur horticulturists to surreptitiously import and propagate species. These uncontrolled introductions are a major avenue for plant immigration, in part because they are difficult to intercept.

Mack (1999) argues for comprehensive standards for evaluating the risks of releasing species, so that benefits can be weighed against potential costs before release. This idea has long formed the basis for evaluating potential biological control agents. For example, the Australian Government has recently (in 1997) introduced a weed risk assessment system that attempts to evaluate risk in new plant introductions. Other countries including the US are considering the adoption of similar systems. Though there are arguments about the reliability of such systems, they at least force a slow-down and a sense of questioning in the hitherto seemingly unrelenting plant introduction process. Currently, around 30% of species are excluded from Australia, where previously very few had been. Formerly, only those known elsewhere as declared weeds were excluded. Mack proposes that testing under quarantine for plants able to persist without cultivation (i.e. watering, nutrients, pest control) would screen out potential invaders. Certainly, plants that had been able to persist for several years without tending were most likely to be weeds amongst those studied by Lonsdale (1994). Unfortunately, these were also the species most likely to be useful in the pastoral situation. Furthermore, such a process would be unlikely to deal well with ornamentals, for which the commercial value of any single species would not be worth the effort of this depth of research.

A Helping Hand

Does increased nitrogen availability favour weed invasion in rangelands?

While rangelands may seem unlikely locations for enhanced nitrogen inputs, fixed forms of nitrogen may be lost to the atmosphere from human-dominated systems and reach ecosystems downwind. Thus, fragmentation and urban expansion are in fact associated with downwind areas of enhanced nitrogen deposition. Because nitrogen is so often limiting to primary productivity, alteration of nitrogen availability has the potential to influence the relative success of non-native plants entering natural communities. *Enhanced nitrogen availability apparently encourages non-native species in semi-arid ecosystems* (Huenneke, 1999). Nitrogen has a significant influence on productivity even in drylands (Lajtha and Schlesinger, 1986; Shachak and Lovett, 1998). In other low-productivity ecosystems, the increase in primary production due to nitrogen additions has enhanced the invasiveness and dominance of non-native species (e.g. Huenneke *et al.*, 1990, for Californian grasslands; Marrs and Lowday, 1992, for heathland; Marrs, 1993). Permanent transects in the Chihuahuan desert of New Mexico, USA, supported significantly greater cover of non-native species after fertilization (Huenneke, unpublished analysis of data from Jornada Long-Term Ecological Research Programme). McIntyre and Lavorel (1994) found that conditions enhancing water availability could encourage the dominance and success of exotics; in semi-arid systems, there can be an interaction of water and nitrogen such that, with additional water, plants can respond to the addition of nitrogen more strongly.

Development and fragmentation

There are increasing pressures for recreation and access to rangelands for multiple use beyond livestock grazing (Huenneke, 1999). *Many of these uses also lead to opportunities for invasive species.* For example, the affordability of water for irrigation has led to expansion of intensive agriculture in some locations within arid or semi-arid regions. The intensity of farming and agricultural practices is positively correlated with the frequency of weedy species (Boutin and Jobin, 1998), and agricultural fields and watering points provide foci from which climatically adapted species might spread (Huenneke and Noble, 1996). Pressures for such expansion of agriculture into lands previously used only for less-intensive grazing will probably increase in many semi-arid areas as human population pressures increase.

Because human transportation facilitates the dispersal of weed propagules and because human development activities create disturbances that promote establishment of weeds, roads and roadside habitats are positively associated with the introduction and spread of weeds. Semi-arid ecosystems and rangelands are not different from other ecosystems in this respect; roadsides are often the locations along which non-natives spread and from which they disperse into surrounding ecosystems.

Thus, it is envisaged that housing developments with their associated gardens, irrigation, fragmentation of rangelands by roads and nitrification from human activities, will work synergistically to accelerate the process of plant invasion.

Can we Increase Weed Control Resources for Rangelands?

Where the free market does not result in the most efficient outcomes, there is market failure, and it may be appropriate for the government to intervene (Pannell, 1999). An example of market failure for rangeland weeds is where a species, for example a pasture grass, spreads from farms into a national park. Unless there is regulation, the farmers creating this 'pollution' will not consider the costs this imposes on others – in this case the taxpayer. (This form of market failure is defined as an externality.) However, theory also indicates that the mere existence of market failure is not a sufficient justification for government involvement. This is because there is a cost to economic efficiency in having the government intervene. For example, it is estimated that the costs of government involvement is generally 40% of the funds collected (Pannell, 1999).

If the government is advised that there will be major external cost if the weed is allowed to spread to the national park, and decides to control the weed, who should pay for the control measures? A commonly cited approach is the 'beneficiary pays' principle. However, it is difficult to identify the true beneficiaries of such control, and it would often end in being defined as the taxpayer. An alternative, the 'polluter pays' principle, would have the farmers pay. The two alternatives will have different economic and social consequences but the decision is a political one, not an economic one.

Pannell briefly reviewed the Australian National Weeds Strategy (Anon., 1999), arguing that its effectiveness is likely to be limited because of the nature of the Australian federation. Responsibility for most laws dealing with the land rests with state governments. Consequently, the National Weeds Strategy would have to be well resourced to buy the cooperation of the states. Unfortunately, this has not been the case, and the Strategy's many worthy goals have been formulated without serious consideration of the mechanisms necessary to achieve them (Pannell, 1999). It is 'large on spirit and low on means'.

In the wake of the Uruguay round of the General Agreement on Tariffs and Trade (GATT), world trade bodies have become increasingly concerned that phytosanitary regulations (such as those governing weed introduction) will be used to subvert free trade agreements. Pannell (1999) warns that, in future, Australia may be challenged by the World Trade Organization to show that the claimed absence of particular weeds from our shores is genuine and worth maintaining through regulation.

Recommended Future Directions

We recommend that for weed management in rangeland ecosystems, we need to take a more strategic approach. After a broad indication of what we mean by 'strategic weed control', we will go on and recommend areas of research and application that will contribute to such an approach, namely:

1. Methods of prioritizing amongst weeds;
2. Integrated weed management;
3. Risk assessment and management; and
4. Quantifying invasibility.

Strategic weed control

The 1990s saw increasing use of a strategic approach in weed control. Possibly the first recorded definition of strategic weed control was a paper in 1988 that drew the distinction very clearly between *strategy*, which is taking a long-term, far-reaching view of how to control weeds, and *tactics*, which includes things like different herbicide mixes, nozzle sizes, which insects to use in biocontrol, etc. (Moody and Mack, 1988).

Increasing emphasis on strategy saw Australia developing its National Weed Strategy during the 1990s. Clearly, a strategy must have scientific and technical underpinnings but it must also be able to command resources in the long term and at a large scale. A truly national strategy must be able to bring about policy and behavioural change, not just increase the exchange of information.

South Africa's 'Working for Water' programme is also strategic in conception – it has long time frames and a national scale, and integrates control measures to achieve its aims (Van Wilgen and van Wyk, 1999). Another manifestation of a strategic approach is the creation of the Cooperative Research Centre (CRC) for Weed Management Systems in Australia, a network of agencies. The three ecosystem-based programmes again reflect a strategic approach, the emphasis being on ecosystems and how to manage them instead of groupings around particular weeds.

The CRC had as its goal the aim of reducing the impact of weeds by 10% by the year 2000. This was difficult to demonstrate for environmental weeds and pasture weeds, not because of any failure to develop and apply control technologies, but because of a shortage of baseline data against which the performance can be measured. It is striking how the basic information that is quite widely available for cropping weeds, such as the distribution of the weed and the relationship between its abundance and economic losses, is little known for weeds in rangelands. There is a great need for this kind of information to guide policy development and resource allocation

Prioritizing amongst weeds

There are too many weeds for us to tackle them all. *Efficient control would be best served if we could prioritize them for control based on their importance, as measured by impact.* One of the authors (M.L.) devised the simple equation, $I = R \times A \times E$, where I is overall impact of an invader, R is range, A is abundance and E is per capita impact. It has been used recently by a number of ecologists to clarify the different components of impact of invaders (e.g. Parker *et al.*, 1999; Williamson, 2001).

Our efforts to quantify impact and prioritize amongst different weed species are likely to founder on the term E and by the error in its measurement (see Parker *et al.*, 1999). We have to face up to this uncertainty. To do an environmental impact assessment, say of a new mine, is very expensive, long term and multidisciplinary. Measuring the per capita impact of an invasive weed is likely to be similarly complex and to do it for all invasive weeds would be too costly.

Instead we must make broad guesses about E based on generalizations from functional groups of plants or invaders. One component of E, which in cropping weeds represents around 30 or 40% of the impact, is the cost of control, and this should be comparatively easy to obtain. We also need to tackle the other parts of the equation above, the range R and the abundance A of species, which should be easier to achieve. Both are easy to measure in comparison with E, simply by surveying; they are an important part of overall impact, and ought to be better known. We need to allocate more resources simply to mapping weeds. This could then be combined with some broad estimations of E for functional groups to help prioritize weeds for control.

Integrated weed management for extensive land systems

A useful description of integrated weed management (IWM), a term increasingly in vogue these days (e.g. Radford *et al.*, 1999), is: 'IWM combines weed control methods in a way that maximizes the impact of control resources while minimizing economic, health, and environmental risks'. It is a form of ecosystem management and it is often referred to as contributing to sustainability. It implies an understanding of the ecology of the weed and of the system it infests, and their interaction with the available control measures, such that the demise of the weed can be achieved with minimal environmental damage and maximal efficiency. Note that, in its widest sense, IWM can even include the policies and regulations enacted to achieve a reduction in weed impact (e.g. Hayes and Gilliam, 1999).

Some elements of an IWM strategy include:

1. A clear vision for the desired ecosystem following the weed's removal and the management regime that ecosystem will be under. This then determines the process of revegetation that should be followed, which in turn allows one to identify the control methods that are incompatible with that process.
2. A battery of control methods (fire, biocontrol, mechanical methods, grazing, competitive pastures, etc.) from which to choose, with an indication of their relative compatibilities. Where biocontrol is involved, the challenge is to integrate living organisms, the biocontrol agents, with other methods that undermine the weed's resource base.
3. A management plan to achieve the desired end, that includes stopping rules or thresholds (see Kriticos *et al.*, 1999; Panetta and

James, 1999) in the event that the control methods prove ineffectual or counter-productive.

The knowledge required for such a strategy may take decades and much funding to generate, and comprehensive IWM may therefore be possible only for a sub-set of our major weeds. The cost of registering a herbicide for use against a particular weed may be prohibitive.

In what may be termed a 'triumph of hope over experience', even now, some workers support the use of introduced species to out-compete and thereby diminish the role of current invaders. Specifically, *Bromus tectorum* has dominated a vast rangeland region in the intermountain west of the US for almost 70 years with little evidence that it is ever replaced. Non-indigenous species, *Kochia prostrata* and *Secale montanum*, have been recommended for introduction to combat *B. tectorum* (Anderson *et al.*, 1990; McArthur *et al.*, 1990). However, although the goal may be to replace a species of little forage value with more valuable forage species, any potential replacement of one non-indigenous species with another seems a dubious practice. Furthermore, neither proposed introduction would totally displace *B. tectorum* (R. Mack, personal communication) The more likely result would be an expanded list of non-indigenous species that must be controlled. Fewer, not more, non-indigenous species in rangelands would seem to be the appropriate goal.

Risk assessment and risk management

We now know that probably the majority of Australia's major weeds were introduced intentionally (Panetta, 1993). For example, in northern Australia, 463 species were introduced over 40 years to improve tropical pasture production. Of these species, 13% became weeds but only 4% became useful; 3% were useful and weedy, so only 1% were useful without being weeds (Lonsdale, 1994). We really should be able to do better than this.

Australia now has a weed risk-assessment system in place. This has been tested and found to be more than 80% accurate in identifying weeds and was adopted as policy in 1997. Other systems of very high accuracy are reported in the recent literature (e.g. Rejmanek and Richardson, 1996; Reichard and Hamilton, 1997). However, this high accuracy seems to fly in the face of ecological theory (Lonsdale and Smith, 2001; Williamson, 2001). Success of an invader is the result of an interaction between chance, genotype and environment. No-one knows how big a contribution each has to success, but environment and chance are certainly very important. How, then, can a screening system that asks questions about the plant only, do such a good job of screening out the weeds? It may be that these estimates of accuracy are overly optimistic. Evaluations of screening systems have included biased sampling, the problem of dropouts, failure to carry out blind evaluation and a failure to take account of the base rate effect (Smith *et al.*, 1999; Lonsdale and Smith, 2001). It is important that we improve weed risk analysis (WRA) theory because, for example, WRA may be used for GMOs or for sleeper weeds, and so the consequences of over-optimism about screening methods may be far-reaching.

Controlling weeds at the point of entry seems like good sense until one remembers that thousands of species enter the country each year, yet only a handful become weeds (in the order of ten or so per year for Australia; Groves, 1998). The transition of species from imported to weedy status involves a series of steps with heavy wastage at each step (e.g. Williamson and Fitter, 1996). Consequently, the number of new weeds emerging from the pool of naturalized species each year depends very little on the rate of importation. In order to reduce the rate of emergence of new weed species using importation controls, it would be necessary to reduce the rate of importation to a trickle (Fig. 8.1). This is a worthy aim but would probably be socially unacceptable.

An alternative view would argue that the weed potential of plant species is effectively unpredictable, and that *we would be better advised to focus resources on risk management than on prediction.* For example, we should be doing much more work on casual and naturalized species – those termed sleeper weeds – because it is from here that the next major weeds will come. Even here, risk analysis theory recommends that when a phenomenon is difficult to predict, it is necessary to put resources into monitoring and risk management. Thus, we might have in place a system of sentinel sites – natural and semi-natural vegetation, close to areas of high human population pressure or agricultural development. These would be monitored repeatedly to alert us when a sleeper weed (defined as an exotic species, casual or naturalized, that has not undergone an increase in abundance to damaging levels) has woken from its slumber. Sentinel groups (e.g. keen local naturalists) might be enlisted to monitor the sites regularly as well as other areas on an *ad hoc* basis for species increasing in abundance. At the same time, we should seek resources to eradicate alien species that are present in small numbers, to reduce the diversity of exotics that in itself increases risk.

In general, a site-based rather than species-based approach will be needed where ecosystems are invaded by mixtures of species.

Invasibility

A further bet-hedging strategy would be to focus on risk assessment of *ecosystems* rather than on risk assessment of *species*. Do different land

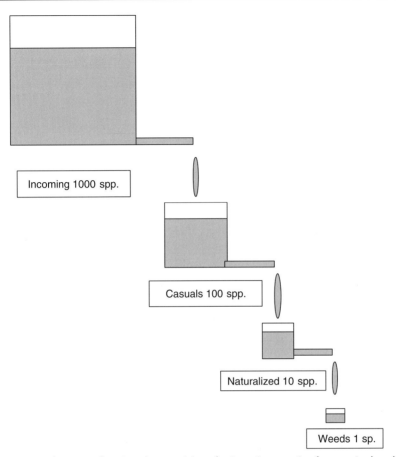

Incoming 1000 spp.

Casuals 100 spp.

Naturalized 10 spp.

Weeds 1 sp.

Fig. 8.1. Schematic showing the transitions for invasive species from arrival to the status of having a negative impact. A series of water tanks represents the pools of species. Each tank drips into the next, and as a rule of thumb we might assume that each tank represents 10% of the volume of the preceding one (see Williamson and Fitter, 1996). Even if we completely halt the transition from incoming to casual – the focus of much effort in quarantine globally – it would take a long time before we will have an effect on the number of weeds. Clearly we should be paying considerable attention to the factors affecting the change from naturalized to weed status.

use types differ significantly in their invasibility – their intrinsic susceptibility to invasion? Past studies of invasibility were flawed because, being based on comparisons of the degree to which different ecosystems were invaded, they confounded invasibility with propagule pressure (Lonsdale, 1999). *We need a new approach that would involve looking at the fates of groups of species introduced into regions over history to compare the success rate* (e.g. proportion of species becoming naturalized, average lag-phase before naturalization, and so on) *of those species in different ecosystems.* This would to some extent control for propagule pressure, so that differences between ecosystems in such variables would be an indication of differences in their invasibilities. Such an approach would also help us in risk assessment because it may be possible to develop generalizations of the relevant risks posed by different sectors to different habitats.

Conclusions

Rangelands globally are amongst the ecosystems worst affected by weeds. The extent of rangelands that are invaded is probably increasing, though monitoring of the problem is poor. Plant invasions result largely from human needs, which result in movement of weed species and degradation of ecosystems so that they become more susceptible to invasion. Increasing pressure of international trade will increase the movement of species around the world, but quarantine barriers to exclude potentially invasive species could be challenged under World Trade Organization procedures unless they can be shown to have a strong scientific basis. On the other hand, agricultural production systems that result in plant invasions without controlling them could also be disputed as representing an unfair trade subsidy to the industry causing the damage. There will be an increasing need for a risk-weighted approach to plant introductions and weed management, and especially an understanding of the costs and benefits of introducing and of excluding proposed introductions. While the battle may be lost against the current generation of weeds, it should be possible to slow the rate of emergence of new weed species, or of new outbreaks by existing weed species, by placing greater emphasis on surveillance, perhaps with a system of sentinel sites.

Acknowledgements

We thank Tony Grice and Ken Hodgkinson for their comments on a draft of the manuscript, and the speakers, authors and participants in the session on 'People and plant invasions of the rangelands' at the VI International Rangeland Congress for their stimulating insights.

References

Anderson, M.R., DePuit, E.J., Abernathy, R.H. and Kleinman, L.H. (1990) Suppression of annual bromegrasses by mountain rye on semiarid mined lands. *Proceedings of a Symposium on Cheatgrass Invasion, Shrub Die-off, and Other Aspects of Shrub Biology and Management.* Forest Service Intermountain

Research Station General Technical Report, INT-276. United States Department of Agriculture, Washington, DC, pp. 47–55.

Anon. (1999) National Weed Strategy (revised edition). AFFA, Canberra.

Babbitt, B. (1998) Statement by Secretary of the Interior, Bruce Babbitt, on Invasive Alien Species, 'Science in Wildland Weed Management' Symposium, Denver, Colorado, April 8, 1998. http://www.nps.gov/plants/alien/press/bbstat.htm

Boutin, C. and Jobin, B. (1998) Intensity of agricultural practices and effects on adjacent habitats. *Ecological Applications* 8, 544–557.

Combellack, J.H. (1989) Resource allocations for future weed control activities. *Proceedings of the 42nd New Zealand Weed and Pest Control Conference*, Auckland, New Zealand, pp. 15–31.

Groves, R.H. (1998) *Recent Incursions of Weeds to Australia 1971–1995*. Cooperative Research Centre for Weed Management Systems Technical Series No. 3. University of Adelaide, Adelaide.

Hayes, D.C. and Gilliam, B.C. (1999) Public policy to prevent spread of noxious weeds by modifying human behaviour. In: Eldridge, D. and Freudenberger, D. (eds) *People of the Rangelands. Building the Future. Proceedings of the VI International Rangeland Congress*. VI International Rangeland Congress Inc., Townsville, Australia, pp. 597–598.

Huenneke, L.F. (1999) A helping hand: facilitation of plant invasions by human activities. In: Eldridge, D. and Freudenberger, D. (eds) *People of the Rangelands. Building the Future. Proceedings of the VI International Rangeland Congress*. VI International Rangeland Congress Inc., Townsville, Australia, pp. 562–566.

Huenneke, L.F. and Noble, I. (1996) Ecosystem function of biodiversity in arid ecosystems. In: Mooney, H.A., Cushman, J.H., Sala, O.E. and Schulze, E.-D. (eds) *Functional Roles of Biodiversity: a Global Perspective*. John Wiley & Sons, Chichester, UK, pp. 99–128.

Huenneke, L.F., Hamburg, S.P., Koide, R., Mooney, H.A. and Vitousek, P.M. (1990) Effects of soil resources on plant invasion and community structure in Californian serpentine grassland. *Ecology* 71, 478–491.

Kriticos, D.J., Brown, J.R., Radford, I.J. and Nicholas, M. (1999) A population model for *Acacia nilotica*: thresholds for management. In: Eldridge, D. and Freudenberger, D. (eds) *People of the Rangelands. Building the Future. Proceedings of the VI International Rangeland Congress*. VI International Rangeland Congress Inc., Townsville, Australia, pp. 599–600.

Lajtha, K. and Schlesinger, W.H. (1986) Plant response to variation in nitrogen availability in desert shrubland community. *Biogeochemistry* 2, 29–37.

Lonsdale, W.M. (1994) Inviting trouble: introduced pasture species in northern Australia. *Australian Journal of Ecology* 19, 345–354.

Lonsdale, W.M. (1999) Concepts and synthesis: global patterns of plant invasions, and the concept of invasibility. *Ecology* 80, 1522–1536.

Lonsdale, W.M. and Smith, C.S. (2001). Evaluating pest screening procedures – insights from epidemiology and ecology. In: Virtue, J. and Groves, R.H. (eds) *Proceedings of a Workshop on Weed Risk Assessment*. CSIRO Publishing, Melbourne, pp. 52–60.

Mack, R.N. (1999) The motivation for importing potentially invasive plant species: a primal urge? In: Eldridge, D. and Freudenberger, D. (eds) *People of the Rangelands. Building the Future. Proceedings of the VI International Rangeland*

Congress. VI International Rangeland Congress Inc., Townsville, Australia, pp. 557–562.

Marrs, R.H. (1993) Soil fertility and nature conservation in Europe: theoretical considerations and practical management solutions. *Advances in Ecological Research* 24, 241–300.

Marrs, R.H. and Lowday, J.E. (1992) Control of bracken and the restoration of heathland. II. Regeneration of the heathland community. *Journal of Applied Ecology* 29, 204–211.

McArthur, E.D., Blauer, A.C. and Stevens, R. (1990) Forage Kochia competition with cheatgrass in central Utah. *Proceedings of a Symposium on Cheatgrass Invasion, Shrub Die-off, and other Aspects of Shrub Biology and Management.* Forest Service Intermountain Research Station General Technical Report, INT-276. United States Department of Agriculture, Washington, DC, pp. 56–65.

McIntyre, S. and Lavorel, S. (1994) Predicting richness of native, rare and exotic plants in response to habitat and disturbance variables across a variegated landscape. *Conservation Biology* 8, 521–531.

Moody, M.E. and Mack, R.N. (1988) Controlling the spread of plant invasions: the importance of nascent foci. *Journal of Applied Ecology* 25, 1009–1021.

Panetta, F.D. (1993) A system of assessing proposed plant introductions for weed potential. *Plant Protection Quarterly* 8, 10–14.

Panetta, F.D. and James, R.F. (1999) Weed control thresholds: a useful concept in natural ecosystems? *Plant Protection Quarterly* 14, pp. 68–76.

Pannell, D. (1999) Paying for weeds: economics and policies for weeds in extensive land systems. In: Eldridge, D. and Freudenberger, D. (eds) *People of the Rangelands. Building the Future. Proceedings of the VI International Rangeland Congress.* VI International Rangeland Congress Inc., Townsville, Australia, pp. 571–576.

Parker, I.M., Simberloff, D., Lonsdale, W.M., Goodell, K., Wonham, M., Kareiva, P., Williamson, M., von Holle, B., Moyle, P., Byers, J. and Goldwasser, L. (1999) Impact: assessing the ecological effects of invaders. *Biological Invasions* 1, 3–19.

Radford, I.J., Kriticos, D., Nicholas, M. and Brown, J.R. (1999) Towards an integrated approach to the management of *Acacia nilotica* in northern Australia. *People of the Rangelands. Building the Future. Proceedings of the VI International Rangeland Congress.* VI International Rangeland Congress Inc., Townsville, Australia, pp. 585–586.

Reichard, S.E. and Hamilton, C.W. (1997) Predicting invasions of woody plants introduced into North America. *Conservation Biology* 11, 193–203.

Rejmanek, M. and Richardson, D.M. (1996) What attributes make some plant species more invasive? *Ecology* 77, 1655–1660.

Shachak, M. and Lovett, G.M. (1998) Atmospheric deposition to a desert ecosystem and its implications for management. *Ecological Applications* 8, 455–463.

Smith, C.S., Lonsdale, W.M. and Fortune, J. (1999) When to ignore advice: invasion predictions and decision theory. *Biological Invasions* 1, 89–96.

Van Wilgen, B. and van Wyk, E. (1999) Invading alien plants in South Africa: impacts and solutions. In: Eldridge, D. and Freudenberger, D. (eds) *People of the Rangelands. Building the Future. Proceedings of the VI International Rangeland Congress.* VI International Rangeland Congress Inc., Townsville, Australia, pp. 566–571.

Williamson, M. (2001) Can the impacts of invasive species be predicted? In: Virtue, J. and Groves, R.H. (eds) *Proceedings of a Workshop on Weed Risk Assessment.* CSIRO Publishing, Melbourne, pp. 20–33.

Williamson, M. and Fitter, A. (1996) The varying success of invaders. *Ecology* 77, 1661–1665.

People and Rangeland Biodiversity

David M.J.S. Bowman

9

Introduction

What ecological factors best distinguish rangeland from agricultural landscapes? One obvious difference is that in rangelands, native vegetation is exploited to support livestock production, whereas in more intensively managed agrosystems the natural ecosystem is virtually replaced. *Indeed, an often unappreciated and undervalued by-product of rangeland grazing has been the conservation of biological diversity.* Thus, a simple answer to the question posed above is the predominance of 'biodiversity': if you substantially eliminate 'biodiversity' from rangelands you end up with agricultural landscapes.

Definitions of Biodiversity

Biodiversity is an easy word to say but a difficult concept to define. The word can be used to describe two related concepts. Biodiversity is a phrase that encapsulates the threat of human-induced mass extinction of species. Alternatively, it can be shorthand for one of the great, unsolved scientific problems: the evolutionary and ecological significance of the enormous diversity of life-forms. The two uses are connected because some biologists believe that by understanding the latter they can contribute to moderating the former. The word carries considerable political freight, as the general public's widespread concern about the current extinction crisis provided political spur for 161

©CAB International 2002. Global Rangelands: Progress and Prospects
(eds A.C. Grice and K.C. Hodgkinson)   117

countries to ratify the International Convention on Biological Diversity. In what follows the word 'biodiversity' is used in both the scientific and the environmental sense.

Perceptions of biodiversity vary with temporal and spatial scales. Some researchers believe that long extinct life-forms, such as the Pleistocene megaherbivores, should be considered when evaluating the health and integrity of rangelands (Flannery, 1999). Historical analyses clearly demonstrate that rangeland biodiversity has changed in response to climatic as well as human influences over the last 1000 years (Thomas, 1999). Even when working in modern-day landscapes, the measurement of biodiversity depends upon temporal perspectives and spatial scale. For instance, Australian experience suggests that the maintenance of biodiversity is dependent upon the subtle variation of biological resources in space and time within vast tracts of superficially homogeneous rangeland (Woinarski, 1999).

The breadth and 'fuzziness' of the concept frustrates comparative analyses and makes operational definitions difficult to formulate. This is well illustrated by the overall increase in the number of plant species and diversity of life-forms and habitats in the South American pampas following transformation of natural grassland to an agricultural landscape (Ghersa and Leon, 1999). Although the original vegetation is now almost completely lost and typically persists only on road verges, there have been few plant extinctions. Conversely, the widespread historical decline of bird species in the superficially 'intact' savannas of northern Australia signals that pervasive (though insidious and incremental) ecosystem decay has followed European colonization of that area (Woinarski, 1999).

It is critical to note that biodiversity is more than creating lists showing the presence or absence of species at particular locations – something akin to preparing the manifest for Noah's Ark. Snapshots of biodiversity may fail to determine the genetic diversity of populations and their long-term viability. Furthermore, a complete appreciation of biodiversity demands consideration of organisms that cannot be seen by the naked eye and which often live below ground, for example symbiotic root fungi (mycorrhizae) and soil mites.

Operational definitions of biodiversity are needed to make this complex concept tractable. Some authors suggest that selected invertebrate species can provide insights into the overall status of rangeland biodiversity and the ecological health of rangeland landscapes. They argue that the advantages of key groups of invertebrates include great abundance, diversity and importance in the functioning of ecological systems (Cook *et al.*, 1999). However, these authors note that the great diversity and abundance of invertebrates is also a disadvantage and that more research is required to identify key indicator species or broader taxonomic groupings. *To be useful for monitoring, even these bio-indicators need to be coupled with landscape attributes that can be identified using aerial photographs and satellite images.*

The concept of biodiversity also incorporates interactions between species, some of which are of fundamental importance to ecosystem function. This is well illustrated by Noble *et al.*'s (1999) hypothesis that the post-European local extinction of the marsupial burrowing bettong may have triggered the loss of biodiversity in *Acacia aneura* shrublands in the Australian arid zone (Fig. 9.1). They posit that recurrent, localized soil disturbance by this marsupial was 'responsible for creating and maintaining . . . a diverse and productive understorey' and associated diversity of soil organisms, much of which is microscopic, taxonomically undescribed and whose ecological function is unknown.

Ecological Services

It is increasingly recognized that these interactions between species provide ecological services, such as the provision of stable soils and reliable supplies of water, upon which rangeland productivity depends. In a time of changing land use and climate, another key ecological

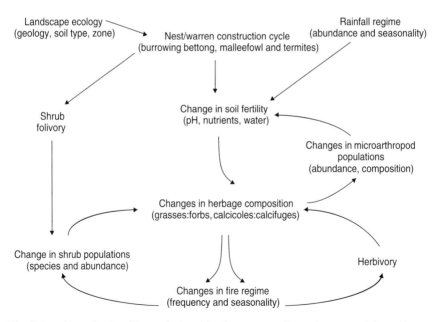

Fig. 9.1. Hypothesized interrelationships between various elements of the soil biota in pre-European *Acacia aneura* shrubland on red earth soils in western New South Wales (Noble *et al.*, 1999).

service provided by biodiversity is to give rangelands resilience to environmental change. Apparently redundant species under one set of environmental conditions may be pivotal for the functioning of rangelands under another set of contingencies. However, it is apparent that the structure, function and response of rangelands to stress and disturbance vary both globally and regionally (e.g. Flannery, 1999), and it is thus difficult to develop general ecological models upon which to base sustainable management. Moreover, the quality and quantity of biological information available to manage rangeland biodiversity also varies enormously. Understanding how rangeland ecosystems work and what role various components of the biodiversity play presents a major research challenge (Fig. 9.2).

For these reasons, the treatment of biodiversity as a purely uni-dimensional and thus readily quantifiable economic resource is most dubious.

Threats

A fundamental driver of the current global extinction crisis is the growth of human populations. In many areas, the level of demand for food and other resources being placed on rangelands are such as to cause environmental degradation and declines in biodiversity (e.g. Bond, 1999; Shourong, 1999). Furthermore, as a result of the continuing urbanization of populations, decisions that determine the fate of rangeland biodiversity are often made by people largely disconnected from the day-to-day realities and ecological constraints of rural

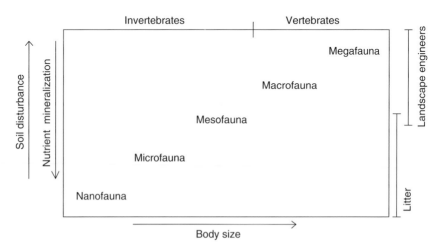

Fig. 9.2. Generalized effect of organisms of different body size on soil structure and soil fertility (Noble *et al.*, 1999).

landscapes. Rangeland biodiversity is also at risk from the destruction of ecosystems during wars (Omar, 1999) that are often themselves a consequence of population pressures and conflict over natural resources.

Probably the single greatest threat to rangelands is their conversion to agricultural or urban land. Rangelands, and thus their biodiversity, have often persisted precisely because the lands are marginal for more intensive agriculture. With population pressure, land-use may be intensified regardless of its marginality, especially in the less-developed world. Land may also be rendered less marginal (and thus more readily convertible to agricultural land) by improved technology such as irrigation, development of new crops and cropping methods, and the successful eradication of diseases afflicting humans and domestic animals (e.g. tsetse fly in Africa). The development of genetically modified organisms (GMOs) has enormous potential to extend this process by allowing the development of crops more tolerant of drought or soil infertility or able to be more efficiently farmed with herbicides and fertilizers. GMOs also present a risk of leakage of genetic material into wild populations, thus threatening rangeland biodiversity. Further, the high efficiency of GMO-based agrosystems is likely to disadvantage populations of wildlife that have been able to 'scrape' a living from the 'waste' of more traditional production systems such as early seed shed of crops and weeds that coexist with crops. It would be a tragic irony indeed if genetic biodiversity were turned against biodiversity in the whole organism and ecological community sense. The development of agricultural systems based on GMOs may also foreclose the opportunity of producing 'clean and green' products from rangelands. There is an increasing demand for such products given widespread consumer antipathy to GMO products. Indeed, biodiversity may 'value add' to rangeland products, hence providing a competitive edge against more traditional agricultural products.

A frequent consequence of human population increases is higher levels of stocking. Many of the world's rangelands have evolved with grazing animals, and biodiversity in some systems may be optimized by moderate grazing pressure (Hart, 1999; Manoharan *et al.*, 1999). Cattle may even be successfully (in an ecological as well as an economic sense) substituted for large wild herbivores (Flannery, 1999). *However, overgrazing is inevitably associated with the loss of biodiversity*, a trend all too apparent in various localities (Ellis *et al.*, 1999; Ferchichi, 1999; Longhi *et al.*, 1999; Manoharan *et al.*, 1999; Reid *et al.*, 1999). Indeed, once human populations exceed the support capacity of livestock populations there is an inevitable shift from range production to more intensive land uses that have negative consequences for rangeland biodiversity (Ellis *et al.*, 1999) (Fig. 9.3).

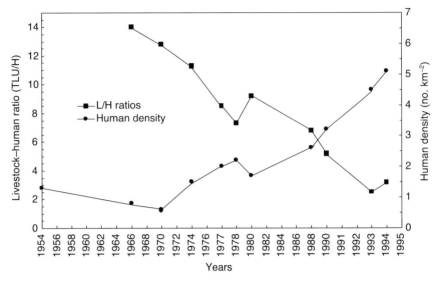

Fig. 9.3.   Declining ratio of livestock populations (measured in tropical livestock units) to human population density in the Ngorongoro Conservation Area, Tanzania. The current ratio of three tropical livestock units per person is the minimum for sustainable rangeland production. Further increases in human population will result in more intensive land-use practices that are incompatible with the stated objective of conserving wildlife populations (Ellis *et al.*, 1999).

Weed invasion is also a major threat to biodiversity. Exotic plants may compete with natives, render habitats unfavourable for wildlife and change ecological processes such as nutrient cycling and fire regimes (Shaw *et al.*, 1999; van Wilgen and van Wyk, 1999). Although exotics can become weeds following unintentional introduction, the vast majority of invasives have been intentionally introduced (Huenneke, 1999). Their spread may be exacerbated by human-caused disturbance.

Strategies for Managing Rangeland Biodiversity

As with commercial fisheries and forestry, the ecologically sustainable management of rangeland, and hence the conservation of biodiversity, demands that there is a balance between wealth creation and the preservation of natural ecosystems. To date, 'sustainable' use of range-lands has often focused narrowly on production rather than overall ecosystem health (Woinarski, 1999). At the most basic level, conservation of rangeland biodiversity is impossible without a widespread acceptance that biodiversity is important – a thought that is still taking root in the rangeland decision-making processes. Indeed, *only recently has the loss of rangeland biodiversity been considered something more than an inevitable 'cost' of economic exploitation* (Woinarski, 1999).

Two contrasting (but potentially complementary) strategies exist for the conservation of rangeland biodiversity. The conventional approach has been the establishment of formally declared conservation reserves. Problems with this approach included a bias against productive landscapes and those landscapes that are devoid of large and/or charismatic animals or other obvious aesthetic values, and a failure to accommodate the rights and traditional practices of indigenous people. *Almost invariably, existing nature reserves fail to adequately conserve a sufficient cross-section of environments.* Individual reserves are often not large enough for animals that migrate across large distances or to buffer ecosystems and species against temporal variation in viability caused by disease, drought, fire and other influences, including the recently recognized threat of climate change.

Because of these problems, many are advocating 'off-reserve' conservation programmes that actively involve local communities. This requires creative approaches such as finding ways to integrate a diversity of uses (e.g. pastoralism, ecotourism, safari hunting, wildlife harvesting, game ranching, harvesting of wild plants for food and medicine, carbon storage and hunting and foraging by indigenous peoples) with biodiversity. If off-reserve management is to be successful, a complex array of factors must be considered. Some of these factors are external, may originate from economically and geographically distant nations, and may be formulated by institutions, such as multinational corporations, that have no real long-term commitment to rangeland. Other factors, such as the accessibility of information and resources to rangeland users and the way they are politically organized, are local. Indeed, off-reserve conservation of biodiversity is impossible without local community control and the explicit incorporation of human needs and aspirations into management planning. In this context it must be accepted that the motivation for community stakeholder involvement is often in response to a range of socio-economic factors rather than a narrow focus on biodiversity conservation. For instance, power and control over natural resources can be seen as a route to achieve political and economic goals pursued by local communities that are not directly related to environmental issues (Fabricius, 1999).

The preservation of rangeland biodiversity requires the commitment of people to balance their immediate self-interest against broader, often intangible, public and ecological values (Fig. 9.4). Critical to the success of this balancing act is a shift from an 'adversarial' approach

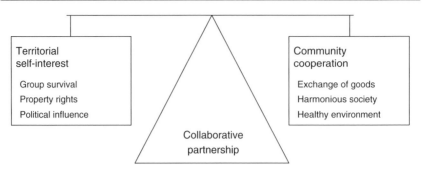

Fig. 9.4. Diagrammatic representation of the sociological factors that must be balanced to achieve collaborative partnerships between individual stakeholders and the broader society (Brunson, 1999).

by management agencies and pressure groups towards land-users to cooperative programmes amongst all stakeholders. There are some encouraging signs that the middle ground can be found to conserve biodiversity in a variety of socio-economic settings (e.g. Barratt, 1999; Brunson, 1999; Fabricius, 1999; Havstad and Coffin Peters, 1999). Clearly such programmes must be economically as well as ecologically sustainable. Much more research is required to understand how to achieve these often disparate goals (James *et al.*, 1999; Perrings and Walker, 1999). In general terms, the economic viability of off-reserve conservation programmes appears to be dependent upon the relationship between land use intensity and biodiversity, and the economic cost and benefits of implementing regimes designed to favour biodiversity. *Although local decision-making is pivotal in off-reserve conservation there is also a clear role for government and non-government interest groups to provide incentives to help local landowners bear the cost of conserving biodiversity in both the developed and developing world.* Acceptable cost-sharing arrangements between 'society' and the individual landholder will probably prove to be central to the success of off-reserve management. This 'incentives' based approach is being accepted in more closely settled areas but needs to be adapted for rangeland. Often funds are only available to support short-term programmes from government or donor programmes (Fabricius, 1999). A good example is the Natural Heritage Trust, recently established by the Commonwealth Government of Australia, that has supported a number of projects designed to conserve biodiversity involving a broad cross-section of stakeholders in Australian rangelands (e.g. Barratt, 1999). A problem with such approaches is that they can be a 'flash in the pan' given the absence of recurrent funding. This is a serious problem given that the response of wildlife and the local community to any management initiative follows idiosyncratic trajectories through time and thus a 'win–win' outcome may take a considerable time to eventuate (Fabricius, 1999) (Fig. 9.5).

There is no doubt that dependence on off-reserve conservation is a risky strategy. Political stability and law enforcement are critical for the success of off-reserve conservation because local communities that make sacrifices for long-term management of biodiversity must be assured that they will be able to reap the rewards. Sadly, globally, there is a strong correlation between political instability and serious threats to biodiversity. In some countries, such as the USA and Australia, systems of legally binding agreements such as covenants and management agreements are being developed to reduce this risk associated with off-reserve conservation. Where they can be implemented, they can be an important means of expanding the biodiversity conservation network. However, economic and social situations can change rapidly, even where there is political stability, thus making it difficult to make off-reserve conservation legally binding. Consequently, some conservation biologists believe that dependence upon off-reserve conservation of biodiversity is far too risky without the back-up of officially designated nature reserves. To be effective such formal reserves should be biogeographically representative, adequately resourced and dedicated in perpetuity (Bond, 1999; Woinarski, 1999). Such reserves can also act as a biodiversity reservoir to stock (or restock) game ranches and degraded landscapes. *Finding an optimal balance between nature reserves and off-reserve conservation strategies demands regional-scale assessment.* Such assessments must identify areas required for conservation reserves and predict, at least in general terms, the consequences

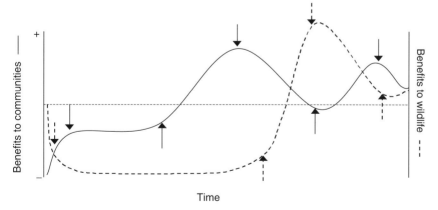

Fig. 9.5. Hypothetical trajectories of positive and negative consequences for wildlife populations and local human communities following a wildlife management initiative. The downward pointing arrows are circumstances that cause a negative response and the upward arrows cause a positive response for either wildlife or humans (Fabricius, 1999).

of various mixes of land uses for biodiversity (Wessels *et al.*, 1999; Woinarski, 1999).

Prognosis for Rangeland Biodiversity

Threats to rangeland biodiversity are in direct proportion to accelerating global economic and environmental change driven by technological innovation and expanding human populations – wittingly and unwittingly, people are determining the fate of the evolutionary heritage contained in rangelands. Rangeland managers are now faced with an extremely complicated juggling act – a balance must be found between conflicting land uses, between short- and long-term economic returns, between management of spectacular and management of ordinary species, and with the rights and responsibilities of indigenous peoples. Further, rangeland managers must be prepared for new challenges and be able to seize opportunities as they arise. Clearly no single strategy can be formulated to 'keep all the balls in the air'. A variety of approaches is required including areas set aside for nature conservation in perpetuity, off-reserve conservation programmes and areas knowingly sacrificed for economic development. The relatively new concept of 'rangeland biodiversity' is important in achieving this balance because it provides a broad planning and policy focus for the management of landscapes outside national parks. Biodiversity should be seen by rangeland professionals as an opportunity to stray from the well-trodden path, to develop new community partnerships and new ways of sustainably managing entire landscapes – there is much exciting and innovative work to be done.

Acknowledgements

I wish to thank all the contributors to the biodiversity session of the VIth International Rangeland Congress for their excellent presentations that formed the basis of this chapter. I am grateful to Don Franklin and John Childs for their thoughtful and helpful comments on an earlier draft.

References

Barratt, R. (1999) Towards best practice management for kowari country. In: Eldridge, D. and Freudenberger, D. (eds) *People of the Rangelands. Building the Future. Proceedings of the VI International Rangeland Congress*. VI International Rangeland Congress Inc., Townsville, Australia, pp. 656–657.

Bond, W.J. (1999) Rangelands and biodiversity in southern Africa – impacts of contrasting management systems. In: Eldridge, D. and Freudenberger, D. (eds) *People of the Rangelands. Building the Future. Proceedings of the VI International Rangeland Congress*. VI International Rangeland Congress Inc., Townsville, Australia, pp. 629–633.

Brunson, M.W. (1999) A sociopolitical model of factors enhancing cross-boundary collaboration. In: Eldridge, D. and Freudenberger, D. (eds) *People of the Rangelands. Building the Future. Proceedings of the VI International Rangeland Congress*. VI International Rangeland Congress Inc., Townsville, Australia, pp. 181–183.

Cook, G.D., Andersen, A.N., Churchill, T.B., Ludwig, J.A., Tongway, D., Williams, R.J. and Woinarski, J.C.Z. (1999) Indicators of ecosystem change in north Australian savannas. In: Eldridge, D. and Freudenberger, D. (eds) *People of the Rangelands. Building the Future. Proceedings of the VI International Rangeland Congress*. VI International Rangeland Congress Inc., Townsville, Australia, pp. 124–125.

Ellis, J., Reid, R., Thornton, P. and Kruska, R. (1999) Population growth and land use change among pastoral people: local processes and continental patterns. In: Eldridge, D. and Freudenberger, D. (eds) *People of the Rangelands. Building the Future. Proceedings of the VI International Rangeland Congress*. VI International Rangeland Congress Inc., Townsville, Australia, pp. 168–169.

Fabricius, C. (1999) Evaluating Eden: who are the winners and losers in community wildlife management? In: Eldridge, D. and Freudenberger, D. (eds) *People of the Rangelands. Building the Future. Proceedings of the VI International Rangeland Congress*. VI International Rangeland Congress Inc., Townsville, Australia, pp. 615–623.

Ferchichi, A. (1999) Rangelands biodiversity in pre-Saharan Tunisia. In: Eldridge, D. and Freudenberger, D. (eds) *People of the Rangelands. Building the Future. Proceedings of the VI International Rangeland Congress*. VI International Rangeland Congress Inc., Townsville, Australia, pp. 666–668.

Flannery, T.F. (1999) Twenty million years of rangelands evolution in Australia and North America. In: Eldridge, D. and Freudenberger, D. (eds) *People of the Rangelands. Building the Future. Proceedings of the VI International Rangeland Congress*. VI International Rangeland Congress Inc., Townsville, Australia, pp. 2–4.

Ghersa, C.M. and Leon, R.J.C. (1999) Landscape changes induced by human activities in the rolling pampas grassland. In: Eldridge, D. and Freudenberger, D. (eds) *People of the Rangelands. Building the Future. Proceedings of the VI International Rangeland Congress*. VI International Rangeland Congress Inc., Townsville, Australia, pp. 624–629.

Hart, R.H. (1999) Cattle grazing intensity and plant biodiversity on shortgrass steppe after 55 years. In: Eldridge, D. and Freudenberger, D. (eds) *People of the Rangelands. Building the Future. Proceedings of the VI International Rangeland Congress*. VI International Rangeland Congress Inc., Townsville, Australia, pp. 646–647.

Havstad, K.M. and Coffin Peters, D.P. (1999) People and rangeland biodiversity – North America. In: Eldridge, D. and Freudenberger, D. (eds) *People of the Rangelands. Building the Future. Proceedings of the VI International Rangeland*

Congress. VI International Rangeland Congress Inc., Townsville, Australia, pp. 634–638.

Huenneke, L.F. (1999) A helping hand: facilitation of plant invasions by human activities. In: Eldridge, D. and Freudenberger, D. (eds) *People of the Rangelands. Building the Future. Proceedings of the VI International Rangeland Congress*. VI International Rangeland Congress Inc., Townsville, Australia, pp. 562–566.

James, C., Stafford Smith, M., Bosma, J., Landsberg, J., Tynan, R. and Maconochie, J. (1999) Production, persistence, economics and extinction: integrating conservation and pastoralism in Australian rangelands. In: Eldridge, D. and Freudenberger, D. (eds) *People of the Rangelands. Building the Future. Proceedings of the VI International Rangeland Congress*. VI International Rangeland Congress Inc., Townsville, Australia, pp. 659–660.

Longhi, F., Pardini, A. and Grammatico Di Tullio, V. (1999) Biodiversity and productivity modifications in the Dhofar rangelands (Southern Sultanate of Oman) due to overgrazing. In: Eldridge, D. and Freudenberger, D. (eds) *People of the Rangelands. Building the Future. Proceedings of the VI International Rangeland Congress*. VI International Rangeland Congress Inc., Townsville, Australia, pp. 664–665.

Manoharan, K., Gobi, M., Shagunthala, J. and Sally, A. (1999) Effect of grazing on biodiversity and soil properties of a semi-arid rangeland at Alwarkurichi, Western Ghats of Tamilnadu, south India. In: Eldridge, D. and Freudenberger, D. (eds) *People of the Rangelands. Building the Future. Proceedings of the VI International Rangeland Congress*. VI International Rangeland Congress Inc., Townsville, Australia, pp. 668–669.

Noble, J.C., Detling, J., Hik, D. and Whitford, W.G. (1999) Soil biodiversity and desertification in *Acacia aneura* woodlands. In: Eldridge, D. and Freudenberger, D. (eds) *People of the Rangelands. Building the Future. Proceedings of the VI International Rangeland Congress*. VI International Rangeland Congress Inc., Townsville, Australia, pp. 108–109.

Omar, S.A.S. (1999) Impact of war on the rangelands of Kuwait. In: Eldridge, D. and Freudenberger, D. (eds) *People of the Rangelands. Building the Future. Proceedings of the VI International Rangeland Congress*. VI International Rangeland Congress Inc., Townsville, Australia, pp. 172–173.

Perrings, C. and Walker, B. (1999) Optimal biodiversity conservation in rangelands. In: Eldridge, D. and Freudenberger, D. (eds) *People of the Rangelands. Building the Future. Proceedings of the VI International Rangeland Congress*. VI International Rangeland Congress Inc., Townsville, Australia, pp. 993–1002.

Reid, R.S., Gardiner, A.J., Kiema, S., Maitima, J.M. and Wilson, C.J. (1999) Impacts of land use on biological diversity in eastern, western and southern Africa. In: Eldridge, D. and Freudenberger, D. (eds) *People of the Rangelands. Building the Future. Proceedings of the VI International Rangeland Congress*. VI International Rangeland Congress Inc., Townsville, Australia, pp. 670–671.

Shaw, N.L., Saab, V.A., Monsen, S.B. and Rich, T.D. (1999) *Bromus tectorum* expansion and biodiversity loss on the Snake River Plain, southern Idaho, USA. In: Eldridge, D. and Freudenberger, D. (eds) *People of the Rangelands. Building the Future. Proceedings of the VI International Rangeland Congress*. VI International Rangeland Congress Inc., Townsville, Australia, pp. 586–588.

Shourong, Z. (1999) Rangeland biodiversity: 'green' foods and medicines from southern China. In: Eldridge, D. and Freudenberger, D. (eds) *People of the Rangelands. Building the Future. Proceedings of the VI International Rangeland Congress*. VI International Rangeland Congress Inc., Townsville, Australia, pp. 653–654.

Thomas, I. (1999) The recent history of rangelands in western Asia. In: Eldridge, D. and Freudenberger, D. (eds) *People of the Rangelands. Building the Future. Proceedings of the VI International Rangeland Congress*. VI International Rangeland Congress Inc., Townsville, Australia, pp. 1048–1052.

van Wilgen, B.W. and van Wyk, E. (1999) Invading alien plants in South Africa: impacts and solutions. In: Eldridge, D. and Freudenberger, D. (eds) *People of the Rangelands. Building the Future. Proceedings of the VI International Rangeland Congress*. VI International Rangeland Congress Inc., Townsville, Australia, pp. 566–571.

Wessels, K.J., Reyers, B. and van Jaarsveld, A.S. (1999) Rangeland transformation in northern South Africa: prospects for biodiversity conservation. In: Eldridge, D. and Freudenberger, D. (eds) *People of the Rangelands. Building the Future. Proceedings of the VI International Rangeland Congress*. VI International Rangeland Congress Inc., Townsville, Australia, pp. 657–658.

Woinarski, J.C.Z. (1999) Prognosis and framework for the conservation of biodiversity in rangelands: building on the north Australian experience. In: Eldridge, D. and Freudenberger, D. (eds) *People of the Rangelands. Building the Future. Proceedings of the VI International Rangeland Congress*. VI International Rangeland Congress Inc., Townsville, Australia, pp. 639–645.

Managing Grazing

10

Mick Quirk

Introduction

Managing grazing to achieve sustainable use of grazing lands has been the primary theme of rangeland management. While there is now increasing emphasis on uses and values of rangelands that are not directly dependent on grazing, the fact remains that grazing, be it from domestic animals, wildlife and/or feral herbivores, is an integral process in practically all rangelands.

The 'traditional' discipline of grazing management was built upon the view that grazing was the primary driver of rangeland condition (e.g. Stoddart *et al.*, 1975; Vallentine, 1990). The evidence of the effects on landscapes of past overstocking, combined with studies of the effects of defoliation on plant morphology, population dynamics and physiology, produced the understanding that grazing was primarily a negative process and therefore required close management. A maxim of grazing management was that pastures should be rested to conserve or replenish root reserves and allow plants to set and distribute seed (Stoddart *et al.*, 1975). Grazing systems were based on either regular deferment of grazing or on periods of rest (varying from weeks to months, depending on system) interspersed with periods of grazing (from days to months, again depending on the particular system). The traditional discipline of grazing management generally took this need for regular rest or deferment as a 'given', and focused its efforts on understanding things like spatial patterns of grazing, grazing distribution, foraging behaviour, diet selection, multi-species grazing, carrying capacity and stocking rate.

In recent years, there have been challenges to the traditional approach to managing grazing. First, *the Clementsian view of rangeland succession* (Clements, 1916) *and its somewhat formulaic application to rangeland management* (e.g. Dyksterhuis, 1949) *has been challenged.* Evidence and opinions were presented to show that, in some situations, grazing was not a primary driver of range condition (e.g. Behnke and Scoones, 1993) and that, in others, some grazing-induced changes in range condition were actually changes in state that were not readily reversible by removal of grazing pressure (e.g. Laycock, 1991). Second, the assumed universal need for some type of grazing system (implying regular rest or deferment, often repeated within the one season or year) has been challenged (e.g. Hart *et al.*, 1993). The majority of studies show no consistent advantage to specialized grazing systems over 'moderate' continuous stocking, all other things being equal. Third, such grazing systems are not applicable to communally managed rangelands, common over most of Africa and central and southern Asia. These rangelands are not fenced and are accessed by many users with different production objectives, resource needs and access rights; these factors make it difficult or impossible to apply a particular grazing system (Hiernaux, 2000).

Where are we then with grazing management? Have we put too much emphasis on grazing as a driver of vegetation change on rangelands? Have we put too much emphasis on grazing systems? When is grazing important? If important, what are the key processes, principles and practices?

Understanding the Effects of Grazing

Are we losing our equilibrium?

The range succession (equilibrium) approach to understanding the dynamics of grazed rangelands has been the basis for traditional grazing management, particularly in the USA and South Africa. This model predicts that grazing pressure, acting against successional tendencies towards a stable single climax, is a primary determinant of range condition; hence, the removal, reduction or appropriate re-timing of grazing pressure should promote a trend towards better condition, as defined by the climax. In terms of population dynamics, this model implicitly assumes a tight coupling between the abundance of herbivores and the productivity and species composition of plants, at least in situations where management interventions (e.g. drought feeding) do not disrupt this coupling.

The recent challenges to this model are based on beliefs, observations and data that are inconsistent with its predictions, for example:

- Grazing-induced changes are not always readily reversed by removal of grazing.
- Multiple steady states, rather than a single climax, are apparent.
- Vegetation changes can be unpredictable, and some may not be caused primarily by grazing.
- Scale affects the nature of the coupling between animal and plant dynamics.

These observations are consistent with a more complex, less predictable view of the dynamics of rangelands, rather than one tied to the notion of a single climax; and one in which the system can change in response to a suite of factors, not just grazing, and may or may not return to its previous condition. Several alternative models have been proposed to help understand and predict this complexity, e.g. state-and-transition (Westoby *et al.*, 1989) and threshold models (Friedel, 1991). Such models attempt to catalogue apparently stable states and the factors that may trigger transitions to other states. State-and-transition has become associated with a non-equilibrium view of range dynamics, even though it does not presume any basis in ecological theory (Westoby *et al.*, 1989). In any case, non-equilibrium concepts (and state-and-transition, by association) have been promoted as being particularly appropriate for inherently variable, arid or semi-arid environments. In these environments, grazing *per se* may not appear to be as important as chance events, and/or grazing effects may be swamped by climatic variability (e.g. Friedel, 1991; Laycock, 1991). State-and-transition type models also better accommodate historical vegetation changes that appear irreversible in many environments, e.g. increases in woody plant densities (e.g. Archer, 1989).

However, non-equilibrium models, in terms of population dynamics, implicitly assume loose coupling between the abundance of herbivores and the productivity and species composition of plants. This implies that, in truly non-equilibrium environments, plant productivity and composition are determined largely by density-independent abiotic factors, such as climate, and that effects due to herbivory are small. Ellis and Swift (1988) argue from their observations in parts of Africa that such non-equilibrium environments exist where livestock numbers appear unrelated to vegetation dynamics. They argue that changes in forage quantity are too large and rapid to be closely tracked by animal numbers, and also that the animal populations experience intermittent die-offs during extended droughts, thus keeping animal densities below levels that would cause degradation. Another factor contributing to the apparent loose coupling of livestock and vegetation in these environments may be the local to regional movement of livestock, generally on a seasonal basis, associated with transhumance. According to this 'new' paradigm, then, livestock grazing is unlikely to be a cause of land degradation in non-equilibrium environments (e.g. see Behnke and Scoones, 1993). According to Ellis (1994), non-equilibrium dynamics are likely to occur in variable, low-rainfall environments.

Thus, a range of views has emerged about the importance of grazing in determining range condition:

1. The traditional 'Clementsian' approach where grazing is a primary determinant, and vegetation responses to grazing management are generally significant and predictable (with implicit assumptions about equilibrium dynamics).

2. The state-and-transition approach which emphasizes triggers to transitions between apparently stable alternative vegetation states, with grazing as only one of the potential triggers (with no pretence to be based on any ecological model and, therefore, no set assumptions about the degree of coupling).

3. The non-equilibrium approach (with assumptions about loose or non-existent coupling) which predicts that, in some environments, grazing is unimportant.

There is clearly some confusion about these approaches, with state-and-transition and non-equilibrium often being considered equivalent, even though the latter is an ecological model and the former a simple catalogue of perceived states and transitions. Recent discussions about the most appropriate model for discerning the importance and likely impact of grazing management have been bogged down due to:

- Lack of testing of the predictions of each model.
- Some lack of interest in the debate from those that actually manage rangelands (or those that advise rangeland managers), which can be typified by the statement: 'It's just an argument amongst ecologists and academics; let's get on with practical management'.

The supposed dichotomy between equilibrium and non-equilibrium systems is mentioned glibly in many papers and reports with little definition about what this means and with little evidence presented to justify the distinctions. The important point is that we want models that help us understand dynamics of rangelands and provide useful predictive capacity for decision-making. Each model should be tested for its contribution to both understanding and predicability, rather than being blindly accepted as the best approach to understanding rangeland dynamics.

Should we mix our models?

How should the contemporary range professional proceed to tease out the importance of grazing in determining range condition? There are several approaches possible. One approach is to look at the rangeland system under consideration and decide whether or not it is likely to behave like the traditional range succession model (e.g. Fernandez-Gimenez and Allen-Diaz, 1999; Walker and Hodgkinson, 1999). Criteria include climate variability and susceptibility to invasion by woody plants. Fernandez-Gimenez and Allen-Diaz (1999) actually collected data (albeit short-term) to test the appropriateness of different models to Mongolian rangeland systems. They found that neither equilibrium nor non-equilibrium models satisfactorily accounted for their observations, and contend that real rangeland systems exist on a equilibrium–non-equilibrium continuum. I contend that *models allowing for gradual, relatively predictable changes in vegetation as well as for sharp transitions and alternative states will be more fruitful than those that exclude either gradual changes or sharp transitions.* Watson *et al.* (1996) recommend that the appropriate model of change in rangeland systems is one which balances the effects of infrequent, unpredictable events with the impacts of small, frequent but cumulative changes.

In northern Australia, and more recently in the USA (see below), a common approach amongst rangeland professionals has been to embrace a so-called non-equilibrium view of life, dismiss the range succession approach completely, and embrace a model like state-and-transition as the most appropriate means of cataloguing observations about vegetation change (e.g. Hall *et al.*, 1994; McIvor and Scanlan, 1994). Observations are usually presented to justify the appointment of likely states and transitions. As mentioned above, there is no general ecological theory involved in this process (even though state-and-transition and non-equilibrium are used interchangeably at times). The approach is about organizing correlative information in such a way as to make it more useful for management purposes (R.M. Rodriguez Iglesias, personal communication).

Interestingly, state-and-transition models generally have grazing as a significant trigger of most transitions; disagreement with Clementsian approaches appears to be about the possibility of alternative 'states', the reversibility of changes in 'state', and the role of chance events and factors such as fire. This implies a strong degree of biotic coupling in these systems (therefore, not truly based on non-equilibrium concepts), but the dynamics are more complex than predicted by the range succession approach. Hence, it seems that the attraction of more complex models relates to considerations of rate of change, reversibility and interaction of several triggers, not any perceived need to embrace the population dynamics implicit in non-equilibrium models.

Embracing state-and-transition is the approach recently adopted by the Natural Resource Conservation Service (formerly the Soil Conservation Service) of USDA in its *National Range and Pasture Handbook* (USDA, 1997). This handbook guides the work of range management extension staff on private rangelands in the USA and, in the past, has been exclusively based on the range succession approach as adapted by Dyksterhuis (1949).

A continuing weakness of this search for the most appropriate model of rangeland dynamics is that most arguments are constructed around data and observations put together to argue a particular case. Apart from a couple of exceptions (e.g. Coppock, 1993; Fernandez-Gimenez and Allen-Diaz, 1999, discussed above), there has been little testing of the applicability of different models. This testing should be rigorous, but true to the intent of the concepts being tested. For example, substantial inter-annual variability in rainfall, resulting in variability in the growth of vegetation, cannot itself be used to contradict the Clementsian model (see Fernandez-Gimenez and Allen-Diaz, 1999). Constancy of climate and not weather was an underlying assumption of Clements' approach. In any case, recent data in the dry tropics show grazing can have somewhat predictable effects on the decline and recovery of range condition, despite large inter-annual variation in rainfall (Ash *et al.*, 1999).

Is grazing ever unimportant?

Are there environments in which non-equilibrium dynamics dominate to the extent that whole landscapes are insensitive to grazing management? The basis of the non-equilibrium approach and its undiscerning application to environments of high rainfall variability, has been challenged by Illius and O'Connor (1999). Illius and O'Connor define equilibrium as strong coupling of animal and plant dynamics, non-equilibrium as weak coupling and disequilibrium as 'equilibrium' in a variable environment (i.e. climatic variation disrupts the stable equilibrium between animals and plants that would be expected to occur under constant conditions). They argue that climatic variation *per se* is insufficient to indicate non-equilibrium environments. Rather, variable environments are at disequilibrium, and the effects of grazing may be magnified (rather than diminished) as such environments have greater variation in grazing pressure than more mesic environments where stable equilibrium conditions are more likely.

Hence, this argument about the relative impact of grazing on rangeland condition is not about whether a more complex model of vegetation dynamics is more appropriate for a particular environment, but rather about whether there is coupling of animal and vegetation dynamics. The degree of inherent coupling would seem to be most easily assessed under natural or semi-natural situations, that is, where there is limited management intervention in animal condition or numbers.

Illius and O'Connor (1999) argue persuasively that it is spatial heterogeneity, rather than climatic variation, that determines the degree and location of coupling between animals and vegetation. This is most clearly apparent when animal numbers are tied to availability of resource-rich areas within their dry-season range (areas close to reliable water). With respect to population dynamics, the animal population is in equilibrium with these 'key resources'. The grazing pressure on other resources in the environment (e.g. wet-season range) will be dependent, therefore, on the relative supply of the key resources. In some cases, the wet-season range may be uncoupled (non-equilibrium in population terms) to the animal population. The unsuspecting observer may then conclude that the landscape as a whole is not likely to be influenced by excessive grazing pressure. Ironically though, the presence of key resources keeps the animal population at higher levels than would occur in their absence and, if the key resources are sufficient to sustain a large animal population, then the outlying range will experience relatively high grazing pressures. Illius and O'Connor (1999) therefore show how landscape heterogeneity may readily explain the apparent loose coupling of animals and vegetation reported from parts of Africa (Ellis and Swift, 1988).

Livestock grazing is likely to induce directional trends in rangeland condition when:

- the grazing landscapes have relatively high amounts of key resources (those resources on which the animal population is dependent during the dry season), such that wet-season range is under relatively high grazing pressures; and/or
- the inherent coupling of animal and plant resources is broken so that animal numbers are artificially inflated above that which can be supported by dry-season resources.

Examples of these situations can be found in the open woodlands of north-eastern Australia. The carrying capacity of most cattle properties in the region is limited by the low nutritive value of the pasture in the dry season. Historically, this has meant that utilization of forage in the wet season (when the perennial grasses are susceptible to overuse) has been relatively low. In more recent years, this coupling of animal population to dry-season resources has been diminished by better adapted cattle and use of supplements, so that wet-season utilization rates have increased with consequent negative effects on land condition

(Quirk *et al.*, 1996). Campbell (1999) describes the evolution of grazing management practices in various rangeland communities of arid and semi-arid Australia, pointing out both the inherent degree of coupling of animal and vegetation dynamics and examples of where this coupling is reduced or enhanced by current management practices.

There are also several cattle properties in north-eastern Australia where key dry-season resources are spatially distinct from the wet-season range, and maintain higher than expected grazing pressures on this range. An example of this occurs where the property, or ranch, includes 'pockets' of grasslands within recent basalt flows. These pockets stay flooded for most of the wet season and are only grazed during the dry season (Rogers *et al.*, 1999). Availability of this high quality dry-season forage allows for much higher carrying capacity than would exist in their absence, such that the wet-season upland areas are under higher than normal grazing pressure.

Thus, *the directional trends due to grazing will depend on the availability of key resources.* These grazing effects may well be intensified, and not diminished, by climatic variability. The general argument for variable semi-arid or arid systems being largely insensitive to grazing appears to be without foundation (see also Borrelli and Oliva, 1999; Holechek *et al.*, 1999). Directional trends due to grazing potentially occur in all rangeland systems, regardless of their conformity to different ecological models, where management interventions inflate livestock numbers consistently above the 'natural' limits of the landscape, as often occurs within sedentary forms of ranching and within some communal systems. Application of 'traditional' rangeland management to parts of Africa has often resulted in unsatisfactory outcomes (e.g. see Behnke and Scoones, 1993). Perhaps this has been due, at least in part, to disruption of the inherent degree of coupling between animal and plant resources (through inappropriate management practices) and not from any inherent insensitivity of the ecosystems to grazing.

Are Grazing Systems Important?

Grazing is clearly an important determinant of rangeland condition, so what is the best way to manage it? In 'natural' environments, herbivore numbers tend to vary in relation to key resources, for example dry-season range, reliable water, etc. Herbivore populations will thus vary over time and space. Nomadic pastoralism also tends towards variable grazing impacts in time and space. The early approach to grazing management, within the range management profession, tried to replicate this apparently natural system of periods of grazing intermingled with periods of rest. To a large extent, this explains the plethora of grazing systems that have been proposed, promoted and/or trialed over the past 80 or so years. In the USA and South Africa, continuous grazing was perceived as the historical cause of rangeland degradation, so some form of interrupted grazing, usually via regular rotation, has been promoted as essential to sound grazing management (e.g. Vallentine, 1990). These systems vary from simple deferment of grazing during a particular season, through rotational resting of paddocks for a month, a season or a year, through to one herd : multiple paddock systems with various time ratios of grazing:rest depending on the objective (accelerated range condition versus high individual animal performance) (see Vallentine, 1990).

Until recently, then, the importance of the grazing system (the spatial and temporal distribution of grazing pressure) was perceived to override any other factor, even stocking rate. This perception was maintained despite empirical evidence to show that grazing systems *per se* had little effect on range condition and often produced lower animal performance compared with continuous grazing at the same overall stocking rate. There have been anecdotal reports of substantial benefits from the commercial adoption of grazing systems (especially the more intensive systems like cell grazing), but most evidence points to these benefits arising from improved control of grazing distribution, better matching of feed demand to feed supply and better animal husbandry generally, rather than from the high frequency of resting, animal impact, or other claimed benefits of the particular system (e.g. Hart *et al.*, 1993; Walker and Hodgkinson, 1999).

The evolution and promotion of grazing systems is not unrelated to the traditional view of how rangeland systems operate, that is, as an equilibrium system. The benefits of rest were perceived to be predictable and reliable, and many of the systems were based on prescription of days grazed and days rested, almost on a calendar basis. Given the rainfall variability inherent in all but the most mesic rangeland environments, such systems now seem naïve and highly risky. These more formulaic grazing systems have left the door open to promoters of other, more flexible systems (e.g. time-control grazing). Such promoters have dismissed much of the research on grazing systems because the systems tested were too prescriptive and calendar driven, rather than being responsive to what was actually happening in the paddocks. However, most evaluations of the more intensive grazing systems over the last 10–15 years have mimicked the flexibility demanded by their promoters (e.g. varying grazing days with pasture growth), and claims of an intrinsic superiority of any particular grazing system are still without credible empirical support.

Is grazing management, then, really just about managing stocking rate? All other things being equal, *stocking rate is the major factor in grazing management* (see Heitschmidt and Taylor, 1991; O'Reagain and Turner, 1992; Walker and Hodgkinson, 1999; Willms *et al.*, 1999).

However, in the real world, things are rarely equal. For example, distribution of grazing is a major problem in most rangelands. Riparian areas and other preferred land types will often be selectively over-utilized regardless of the overall stocking rate (see Stuth, 1991). As pointed out by Walker and Hodgkinson, many grazing problems are a function of poor distribution of grazing and the reported benefits of grazing systems for commercial ranches can often be linked to positive effects on grazing distribution. Other factors which can enhance grazing management greatly are species of livestock and the season/ timing of grazing (Walker and Hodgkinson, 1999). Management of stocking rate through time is another key factor, given the inter-annual variability in forage production that characterizes most rangelands (O'Reagain and Bushell, 1999).

Given that distribution of grazing is relatively even, that the appropriate mix of livestock species is used and that grazing is avoided when the forage species are particularly susceptible (e.g. high altitude temperate country at the start of the growing season), what else can we do to improve either range condition (and hence pasture production) and/or harvest efficiency? While formulaic resting of rangeland via formal grazing systems generally seems an ineffective, or at least inefficient, way of improving range condition, it appears that resting rangelands at carefully targeted times, alternating with periods of 'moderate' use, can both greatly accelerate improvements in range condition and, perhaps, allow higher overall utilization levels. Targeted spelling (termed tactical grazing by Hacker et al., 1999; see other examples from Friend et al., 1999; Johnson and Hodgkinson, 1999) seeks to exploit an understanding of the interactions between intensity/timing of grazing and seasonal conditions as the basis for planning strategic rest to maintain/improve range condition. Intensive early stocking of mixed and tallgrass prairies in the southern USA (Gillen and Sims, 1999) is another example of a strategic system which demonstrably improves pasture and animal production in a sustainable manner. Similarly, early wet-season spelling in dry tropical savannas hastens recovery of range condition and appears to increase the tolerable level of annual utilization from 25% to 50% (Ash, 1998). This suggests that sustainable grazing strategies could be developed which allow periods of relatively high utilization interspersed with periods of rest at critical times. This could be an important basis for more sustainable grazing systems in many rangeland environments, especially those in which the short-term productivity of herds is optimized at utilization levels that may be inconsistent with long-term sustainability.

Ironically then, given the growing disillusionment with traditional grazing systems based on regular and/or frequent rotation, there appears much to be gained from combining the basics of sound grazing management (see Walker and Hodgkinson, 1999) with flexible, ecologically based grazing strategies that accelerate improvements in range condition and, perhaps, allow for modest increases in overall grazing pressure.

The opportunities and constraints for managing grazing discussed so far are largely those associated with sedentary ranching systems using individually owned or leased land. A somewhat different array of opportunities and constraints is associated with nomadic livestock grazing and communal access to land, even though the principles underlying grazing management are the same. Transhumance and nomadic systems seek optimal use of pastoral resources over large areas through tracking of resource availability; these systems are increasingly hampered by either expansion of cropping areas and/or increasing population (Hiernaux, 2000). With communally owned land, development of effective grazing systems requires collective action to reach agreement within the community on the management of grazing pressure, on the demarcation of rangeland units and on interaction with neighbouring communities and visiting pastoralists (Hiernaux, 2000).

Conclusions

Grazing will continue to be an important process in all rangelands, regardless of their primary use, and managing grazing will continue to preoccupy landholders and others interested in the sustainable and productive use of rangelands. Managing grazing is obviously more than understanding biophysical processes and relationships: there is a socio-economic context which interacts strongly with the biophysical processes to determine current practices, perceptions and prospects in relation to managing grazing (e.g. see Borrelli and Oliva, 1999; Campbell, 1999; de Leeuw *et al.*, 1999; Vetter and Bond, 1999; Willms *et al.*, 1999). Be this as it may, significant contributions to sustainability and productivity have resulted from regional research and extension programmes which seek a better understanding of grazing ecology, an appreciation of the evolution of current grazing practices, links between grazing management and productivity, and information and tools which assist decision-making (see Borrelli and Oliva, 1999; Willms *et al.*, 1999).

Future progress in managing grazing will be based on a more general appreciation and understanding of vegetation dynamics in relation to spatial heterogeneity (rather than the mix of confusion and disinterest which currently exists), more effective communication and demonstration of the benefits that arise from following the basic principles of sound grazing management, and a more holistic approach to grazing management research and extension which does not isolate grazing from other parts of the enterprise or community.

Other critical areas of research and development include:

- Relationships between spatial heterogeneity, range condition, grazing strategies, livestock production and profitability.
- Relationships between landscape function, vegetation structure and composition, and biodiversity.
- Development of ecologically based grazing strategies and options that go beyond the traditional range management approaches (as represented by 'conservative' stocking and 'formal' grazing systems).
- Identification of the policy and infrastructure requirements that can assist individuals and/or communities to pursue sustainable grazing practices.

Acknowledgements

I thank Peter O'Reagain, Andrew Ash, Pierre Hiernaux and Tony Grice for helpful comments.

References

Archer, S. (1989) Have southern Texan savannas been converted to woodlands in recent history. *American Naturalist* 134, 545–561.

Ash, A.J. (1998) Managing woodlands: developing sustainable beef production systems for northern Australia. In: Walker, B. and Lambert, J. (eds) *The North Australia Program: 1998 Review of Improving Resource Management Projects.* Meat and Livestock Australia, Sydney, Australia, pp. 9–14.

Ash, A.J., Corfield, J.P. and Brown, J.R. (1999) Patterns and processes in loss and recovery of perennial grasses in grazed woodlands of semi-arid tropical Australia. In: Eldridge, D. and Freudenberger, D. (eds) *People of the Rangelands. Building the Future. Proceedings of the VI International Rangeland Congress.* VI International Rangeland Congress Inc., Townsville, Australia, pp. 229–230.

Behnke, R.H. and Scoones, I. (1993) Rethinking range ecology: implications for rangeland management in Africa. In: Behnke, R.H., Scoones, I. and Kerven, C. (eds) *Range Ecology at Disequilibrium.* Overseas Development Institute, London, pp. 1–30.

Borrelli, P. and Oliva, G. (1999) Managing grazing: experiences from Patagonia. In: Eldridge, D. and Freudenberger, D. (eds) *People of the Rangelands. Building the Future. Proceedings of the VI International Rangeland Congress.* VI International Rangeland Congress Inc., Townsville, Australia, pp. 441–447.

Campbell, G. (1999) Managing grazing: a perspective from the semi-arid rangelands of Australia. In: Eldridge, D. and Freudenberger, D. (eds) *People of the Rangelands. Building the Future. Proceedings of the VI International Rangeland Congress.* VI International Rangeland Congress Inc., Townsville, Australia, pp. 437–440.

Clements, F.E. (1916) *Plant Succession: an Analysis of the Development of Vegetation.* Carnegie Institution Publication 242, Washington, DC.

Coppock, D.L. (1993) Vegetation and pastoral dynamics in the southern Ethiopian rangelands: implications for theory and management. In: Behnke, R.H., Scoones, I. and Kerven, C. (eds) *Range Ecology at Disequilibrium.* Overseas Development Institute, London, pp. 42–61.

de Leeuw, P.N., Hiernaux, P. and Miheso, V.M. (1999) Grazing pressure in sub-Saharan Africa: an historical perspective. In: Eldridge, D. and Freudenberger, D. (eds) *People of the Rangelands. Building the Future. Proceedings of the VI International Rangeland Congress.* VI International Rangeland Congress Inc., Townsville, Australia, pp. 551–553.

Dyksterhuis, E.J. (1949) Condition and management of rangeland based on quantitative ecology. *Journal of Range Management* 2, 104–115.

Ellis, J.E. (1994) Climate variability and complex ecosystem dynamics: implications for pastoral development. In: Scoones, I. (ed.) *Living with Uncertainty.* Intermediate Technology Publications, London, pp. 37–46.

Ellis, J.E. and Swift, D.M. (1988) Stability of African pastoral ecosystems: alternate paradigms and implications for development. *Journal of Range Management* 41, 450–459.

Fernandez-Gimenez, M.E. and Allen-Diaz, B. (1999) Testing non-equilibrium models of rangeland vegetation dynamics in Mongolia. *Journal of Applied Ecology* 36, 871–885.

Friedel, M.M. (1991) Range condition assessment and the concept of thresholds: a viewpoint. *Journal of Range Management* 44, 422–426.

Friend, D.A., Dolan, P.L. and Hurst, A.M. (1999) Effect of spelling on the growth of a native grass pasture in Tasmania, Australia. In: Eldridge, D. and Freudenberger, D. (eds) *People of the Rangelands. Building the Future. Proceedings of the VI International Rangeland Congress.* VI International Rangeland Congress Inc., Townsville, Australia, pp. 489–490.

Gillen, R.L. and Sims, P.L. (1999) Intensive early stocking of southern mixed grass prairie, USA. In: Eldridge, D. and Freudenberger, D. (eds) *People of the Rangelands. Building the Future. Proceedings of the VI International Rangeland Congress.* VI International Rangeland Congress Inc., Townsville, Australia, pp. 229–230.

Hacker, R.B., Hodgkinson, K.C., Bean, J. and Melville, C.J. (1999) Response to intensity and duration of grazing in a semi-arid wooded grassland. In: Eldridge, D. and Freudenberger, D. (eds) *People of the Rangelands. Building the Future. Proceedings of the VI International Rangeland Congress.* VI International Rangeland Congress Inc., Townsville, Australia, pp. 458–459.

Hall, T.J., Filet, P.G., Banks, B. and Silcock, R.G. (1994) State and transition models for rangelands, 11. A state and transition model for Aristida-Bothriochloa pasture community of central and southern Queensland. *Tropical Grasslands* 28, 270–273.

Hart, R.H., Bissio, J., Samuel, M.J. and Waggoner, J.W., Jr (1993) Grazing systems, pasture size, and cattle grazing behaviour, distribution and gains. *Journal of Range Management* 46, 81–87.

Heitschmidt, R.K. and Taylor, C.A. (1991) Livestock production. In: Heitschmidt, R.K. and Stuth, J.W. (eds) *Grazing Management: an Ecological Perspective.* Timber Press, Portland, Oregon, pp. 161–178.

Hiernaux, P. (2000) Implications of the 'new rangeland paradigm' for natural resource management, In: Adriansen, H., Reenberg, A. and Nielsen, I. (eds) *The Sahel: Need for Revised Development Strategies*. SEREIN Occasional Paper No. 11, pp. 113–142.

Holechek, J.L., Thomas, M., Molinar, F. and Golt, D. (1999) Stocking desert rangelands: what we've learned. *Rangelands* 21, 8–12.

Illius, A.W. and O'Connor, T.G. (1999) When is grazing a major determinant of rangeland condition and productivity? In: Eldridge, D. and Freudenberger, D. (eds) *People of the Rangelands. Building the Future. Proceedings of the VI International Rangeland Congress*. VI International Rangeland Congress Inc., Townsville, Australia, pp. 419–424.

Johnson, P.S. and Hodgkinson, K.C. (1999) Tactical grazing for perennial grass survival – an Australian vs USA comparison. In: Eldridge, D. and Freudenberger, D. (eds) *People of the Rangelands. Building the Future. Proceedings of the VI International Rangeland Congress*. VI International Rangeland Congress Inc., Townsville, Australia, pp. 474–476.

Laycock, W.A. (1991) Stable states and thresholds of range condition on North American rangelands: a viewpoint. *Journal of Range Management* 44, 427–433.

McIvor, J.G. and Scanlan, J.C. (1994) State and transition model for the northern spear grass zone. *Tropical Grasslands* 28, 256–259.

O'Reagain, P. and Bushell, J. (1999) Testing grazing strategies for the seasonably-variable tropical savannas. In: Eldridge, D. and Freudenberger, D. (eds) *People of the Rangelands. Building the Future. Proceedings of the VI International Rangeland Congress*. VI International Rangeland Congress Inc., Townsville, Australia, pp. 485–486.

O'Reagain, P.J. and Turner, J.R. (1992) An evaluation of the empirical basis for grazing management recommendations for rangeland in southern Africa. *Journal of the Grassland Society of Southern Africa* 9, 38–49.

Quirk, M.F., Ash, A.J. and McKillop, G. (1996) A case study of Dalrymple Shire, North-East Queensland. In: Abel, N. and Ryan, S. (eds) *Sustainable Habitation in the Rangelands*. CSIRO, Canberra, pp. 71–83.

Rogers, L.G., Cannon, M.G. and Barry, E.V. (1999) *Land Resources of the Dalrymple Shire*. Queensland Department of Natural Resources, Brisbane.

Stoddart, L.A., Smith, A.D. and Box, T.W. (1975) *Range Management*. Hill Book Co., Montreal.

Stuth, J.W. (1991) Foraging behaviour. In: Heitschmidt, R.K. and Stuth, J.W. (eds) *Grazing Management: an Ecological Perspective*. Timber Press, Portland, pp. 65–83.

USDA (1997) *National Range and Pasture Handbook*. Natural Resources Conservation Service, Washington, DC.

Vallentine, J.F. (1990) *Grazing Management*. Academic Press, San Diego, California.

Vetter, S. and Bond, J.W. (1999) What are the costs of environmental degradation to communal livestock farmers? In: Eldridge, D. and Freudenberger, D. (eds) *People of the Rangelands. Building the Future. Proceedings of the VI International Rangeland Congress*. VI International Rangeland Congress Inc., Townsville, Australia, pp. 537–538.

Walker, J.W. and Hodgkinson, K.C. (1999) Grazing management: new technologies for old problems. In: Eldridge, D. and Freudenberger, D. (eds) *People of the Rangelands. Building the Future. Proceedings of the VI International Rangeland*

Congress. VI International Rangeland Congress Inc., Townsville, Australia, pp. 424–430.

Watson, J.W., Burnside, D.G. and Holm, A. McR. (1996) Event-driven or continuous: which is the better model for managers? *The Rangeland Journal* 18, 351–369.

Westoby, M., Walker, B. and Noy-Meir, I. (1989) Opportunistic management for rangelands not at equilibrium. *Journal of Range Management* 42, 266–274.

Willms, W.D., Abouguendia, Z.M. and Freeze, B. (1999) Managing the Canadian mixed prairie for improved productivity. In: Eldridge, D. and Freudenberger, D. (eds) *People of the Rangelands. Building the Future. Proceedings of the VI International Rangeland Congress.* VI International Rangeland Congress Inc., Townsville, Australia, pp. 431–437.

Rehabilitation of Mined Surfaces

11

Gerald E. Schuman and Edward F. Redente

Introduction

Surface mining disturbs tens of thousands of hectares of the world's rangelands each year. Mining of such resources as aggregate, bauxite, clay, coal, iron, copper, gold and numerous other minerals results in drastic localized disturbance of these ecosystems. Rangelands of the world play many roles, including producing forage for domestic livestock, as water catchments, providing forage and habitat for wildlife and offering recreation opportunities and general aesthetic value. Rangelands contribute significantly to national parks, wildlife reserves, vegetation reserves and as lands for indigenous people in many parts of the world. Therefore, rehabilitation of the land surface after mining is an important process for both the rangeland and its people.

Over the past two to three decades, the science of mined land rehabilitation has been advanced significantly in the Western world. In the early 1970s mined land rehabilitation was concerned with stabilizing the soil and landscape, mainly by establishing vegetation. Knowledge of land rehabilitation is now well advanced. *Rehabilitation science is now concerned with establishing diverse plant communities, especially using native species, and assessing the process.* The questions now being addressed are: (i) What are the criteria for determining land rehabilitation success and how can it easily be measured? (ii) How can post-revegetation management (grazing, burning, mowing, fertilization) be used to direct plant community development, hence rehabilitation? Evaluation of rehabilitation success must also involve

©CAB *International* 2002. *Global Rangelands: Progress and Prospects* (eds A.C. Grice and K.C. Hodgkinson)

economic and social considerations. Compared with biological factors these are much more difficult to quantify because social/aesthetics issues are personal and will differ widely within and between societies. Assessing the social aspects of rehabilitation success will be difficult. In summary, many interesting challenges still await scientists involved in ensuring that rehabilitation of mined lands meets ecological, economic and social criteria of sustainability.

In many developing countries rehabilitation of mined land is still in its infancy. This is the result of economic factors: resources are limited and priorities lie elsewhere. In many Asian, Eastern European, African, Middle East and South American countries land rehabilitation is not a priority because with limited financial resources the basic needs of people are deemed more important than the need to rehabilitate disturbed or degraded lands. However, in many of these countries, mined land rehabilitation is becoming an issue because of the environmental degradation associated with mining and the increasing demand for water and land resources. *Developing countries are devising mined land rehabilitation programmes but without the underpinning research because of lack of finance. Technology transfer from developed countries is a critical need* if developing countries are to protect their land resources and return mined lands to sustainable production. Countries that have developed successful land rehabilitation technologies for use after mining must work with scientists, industries and governments to promote its application in developing countries.

Countries that have developed knowledge about the processes for stabilizing mined landscapes and restoring productivity to mined lands are now addressing several outstanding issues. Knowledge is needed about the use of native plant species, post-revegetation management and criteria for successful rehabilitation. These issues are really refinements in understanding rehabilitation but are needed by mining companies to fully meet the expectations of local people and the broader community.

Use of Native Species

Plant species native to mined areas are increasingly being used in rehabilitation and in many mines they are required under legislation. Exotic plant species may be used if the area is to be managed for improved pasture or if conditions have been identified that would limit revegetation success with native species. *Revegetation with native species generally leads to a more diverse plant community, which is highly desired or required. In contrast, the use of exotic species in mined land rehabilitation leads to lower diversity* because of the highly competitive nature of many of the exotic plant species that are commonly used. Use of native species also results in more self-sustaining systems because the selection process for exotic cultivars generally results in genotypes with requirements for high soil fertility and rangeland soils are commonly low in nutrients. Plant communities established with native species more effectively meet the demands of multiple uses and blend more naturally with the surrounding landscapes. Revegetation with native species on grazing lands generally allows for more uniform management practices of the rehabilitated lands along with those of the surrounding undisturbed native lands; whereas lands rehabilitated using exotic species require very specific management practices that may be quite different from those required by native rangelands. Bellairs and Davidson (1999) suggest that the establishment of native plant species be assessed through a hierarchy of functional processes: stability, hydrology, propagules, seed germination, soil toxicity and deficiency, competition, and nutrient cycling. The more stable the landform (less runoff and erosion) the better the potential for plant establishment. Once 50% ground cover is achieved the landscape becomes resistant to erosion (Bellairs and Davidson, 1999).

Poor water infiltration is a key limiting factor in successful plant establishment on mined surfaces. In areas of limited precipitation, ripping/scarification and other micro-climate and water harvesting methods improve the chance of successful seedling establishment. For tree species, moonscapes (0.25 ha ponds/depressions) and water ponding banks will concentrate water and store it deeper in the soil. The main source of seeds of native species are the soil seed-bank, collected seed, seed associated with mulch, windborne seed, and seed dispersed by fauna. Research in Australia shows that the seed in native vegetation mulch provides up to about 90% of the species establishing on mined surfaces, the soil seed-bank about 10% and the collected and broadcast seed only about 1%. The seed of native species is often dormant though the mechanism of dormancy differs among species. They may have physical and/or physiological dormancy mechanisms and may require mechanical, temperature and/or moisture treatment to overcome dormancy. In Australia, some native species respond very positively to treatment with plant-derived smoke.

Low soil fertility and potentially toxic soil characteristics also significantly impede native plant establishment and should be considered in selecting seed species mixtures. In saline–sodic soils or spoils, salt-tolerant species will be required. High levels of fertility can result in significant competition between weedy species and the native perennial species during establishment. Competition from weedy species and exotic species used for temporary ground cover can create stresses that limit or hinder establishment of the desired native species. Management considerations should be taken into account early in the rehabilitation process to ensure that plant species selected for the rehabilitated plant community are compatible with planned management options and unplanned activities such as fire. Does the undercover in shrub and tree communities represent a fire hazard that will limit the success of establishing these species? Are the species resilient to grazing?

Cultural practices also affect the development of the plant community (Redente and Keammerer, 1999). Plant establishment is very dependent upon soil moisture conditions during the first year. Method of seeding (drill versus broadcast) appears to have little effect on establishment, with soil factors playing a greater role in plant community structure development than the mixture of native and exotic species utilized.

In general, the use of native species for rehabilitation of mined surfaces in the USA has increased over the last two decades. Legal requirements, the belief that natives are better and more appropriate, and a better understanding of the role of native species in re-establishing biological diversity for enhanced wildlife and recreational value are reasons behind this shift. Native plant communities have generally been found to be more self-sustaining and require less management inputs. To ensure greater success in re-establishing native species on disturbed rangelands, whether degraded by mining, over-grazing, severe drought or repeated fires, requires improved knowledge of seed-bed ecology, seed quality and seed dormancy. Development of a native seed production industry would also greatly enhance the use of native species because of better seed availability, dependability and reduced prices. However, in many instances, exotic species lead to a better outcome and should be used. For example, if mined surfaces are very prone to erosion and stabilizing these lands is critical to prevent the loss of the soil resource, then exotics may be desirable because they are more easily and quickly established than native species. Here, mixtures of native and easily established exotic species are more appropriate. Native species that are more easily established and exotic species that do not strongly compete can be used. Where surfaces are to be used for agricultural production it is often appropriate to establish highly productive exotic species.

Management of Rehabilitated Areas

Management of grazing and fire is important for enhancing mine land rehabilitation. As stated earlier, if grazing is the post-mine land use, it must be demonstrated that the rehabilitated lands can support livestock grazing before the mining company can be relieved of rehabilitation liability. Vicklund (1999) demonstrated that mined lands in north-eastern Wyoming (USA) are capable of sustaining livestock grazing. The average weight gain by heifers was 0.8 kg day^{-1} over a 15-year period, which was greater than that of heifers grazing adjacent native rangeland. Plant species diversity also increased over this 15-year period relative to that of the original seed mixture sown. Areas seeded in 1978 had seven additional C_3 grass species and several shrub and forb species by 1997. Cattle grazing can be used to modify the plant community by rotating livestock more rapidly when desirable species are actively growing and more slowly when less desirable species are actively growing. Change in intensity of grazing pressure can selectively impose more stress on some species and by doing so, alter the composition of the regenerating plant community.

Fire will also reduce or eliminate competition from undesirable species. Wet-season burning in northern Australia reduces fuel loads and alters the seed-bank of undesirable species (Williams and Lane, 1999). Here, wet-season burning prior to removal of topsoil reduces or eliminates seed reserves of the annual *Sorghum* spp. This species competes strongly with more desirable native species especially during the establishment phase. Wet-season burning also reduces fuel loads that might otherwise result in more intense, late dry-season fires that would be detrimental to more desired species, especially shrub and tree species. Williams and Lane (1999) propose that rehabilitating plant communities be burned every second season to control fuel load and the *Sorghum* seed-bank. Burning enhances plant community vigour and reduces the density of shrub species. This enhances forage production and habitat for wildlife. Both mowing and burning in the early stages of rehabilitation reduces competition from annual weeds that typically occurs in the first 1–3 years after reseeding.

Evaluating Rehabilitation Success

The success of rehabilitation must be considered in the context of regulatory, utilitarian and ecological approaches. These approaches overlap and have some common points.

The US federal and state regulatory approach to assessing the success of rehabilitation in the Great Plains and Intermountain West regions has many benefits as well as problems (Giurgevich, 1999). Specific rehabilitation standards are a part of the mining permit process and are enforced by State and/or Federal entities. Standards are not imposed at the end of the mining and rehabilitation process, but are incorporated into the pre-mine, permit, review and approval process. Inspection and evaluation of the success of the rehabilitation are integral parts of the mining process. To ensure that rehabilitation is satisfactorily completed, financial bonding of the mining company to the public entity(s) is required. The level of bonding is determined by the public entity and is retained by them until satisfactory rehabilitation has been achieved. The rehabilitation process must take into account all land-use capabilities and disturbed lands must be restored to a standard that is at least equivalent to that of the pre-mining situation.

Another approach, and one that relies on subjective and non-statistical evaluation, is sometimes referred to as the 'utilitarian approach' to assessing rehabilitation. This is mainly a qualitative assessment where resource stability and post-mine land use are key measures of success. If the lands are to be grazed by livestock, then a post-mine grazing programme is necessary to document capability of the land to support grazing. For example, in the USA, a requirement for rehabilitation of rangelands is that cover and production must be at least 90% of the pre-mine levels. The re-established plant cover, production and diversity must be achieved in the last 2 years of the bonding period and the species established must have the same seasonal dynamics as those in the pre-mine situation. However, there have been several examples showing that a 2–3 species mixture of cool-season grasses will provide season-long grazing for livestock without any warm-season species present. Therefore, if the post-mine use of the land is to be for livestock grazing only, there is no need to establish a warm-season component in the rehabilitation.

Evaluation of species diversity is not easy and the use of standard methods or the 'reference area' approach are not feasible. In fact, numerous forums on the subject of species diversity evaluation have been held over the past 2 years. There is also some interest in using trends in community development as a way of assessing species diversity.

Shrub re-establishment requirements vary but the densities of shrubs required are generally much lower than in the pre-mine setting. Wyoming (USA) has a specific shrub density requirement that states that 20% of the reclaimed land area must have 1 shrub m^{-2} in a mosaic pattern. Much debate on shrub re-establishment requirements has occurred over the past decade because of the difficulty of re-establishing native shrub species.

Other major concerns deal with restoration of the original land contour. This has basically been redefined to mean that the landscape has erosionally stable post-mine drainage channels whose patterns are similar to pre-mine patterns. The resulting topography is generally a gently rolling mixture of uplands and drainage bottomlands. Groundwater issues are also a controversial topic. How do we assess the effect of mining on the groundwater in a 10–20-year bonding period when groundwater models predict that restoration may take hundreds of years?

A more basic ecological approach to assessing mine land rehabilitation success has been proposed in Australia (Tongway and Murphy, 1999). This approach describes the use of indicators of ecological development and stability. Tongway and Murphy (1999) believe that the indicators should be derived within an adaptive framework, a learning process within a specific biophysical and socio-economic context. Their conceptual framework for assessing rehabilitation success recognizes landscape function in terms of: (i) spatial connectivity between ecosystem components including feedback mechanisms; (ii) importance of ecosystem processes in long-term functioning (sustainability); and (iii) the 'economy' of vital resources. They propose that field measurements rely on spatial expression (evaluation of underlying processes rather than specific biota and on soil surface classes). The key features of this approach are to assess resistance to erosion, infiltration and nutrient cycling status over time using permanent transects and comparing these data to analogue or reference sites. The analogue sites provide a context but Tongway and Murphy (1999), like Giurgevich, are careful to state that the rehabilitation site should not be expected to have the same 'indicator' value as the analogue site. Indicator evaluations can be arranged into discrete stages of ecosystem development and 'indicator' value ranges are developed for the various phases of rehabilitation development.

Rehabilitation standards in South Africa consider topography, soils, vegetation, land capability class and post-mining land use (Rethman *et al.*, 1999). They have established guidelines for maximum slopes based on land use such as arable lands (10–12%) and grasslands (12–18%); however, progressive mining companies recommend 7–8% maximum slopes. Topography most suitable to agricultural production is desired but creative landscaping is permitted to provide a variety of habitats to serve to increase diversity of fauna and flora, for game conservation and recreation. Soil is identified as a key element in successful rehabilitation and because of the desirable characteristics of topsoil (i.e. organic matter, seed-bank and fertility) substitute material is only allowed if the soil is totally undesirable because of salinity or other physical/chemical characteristics. Revegetation has been successful with heavy seeding rates to stabilize the soil. Productivity has been demonstrated to be equal to or greater than that of unmined areas on lands of similar potential. Most of the revegetation is to perennial pastures; however, fertility needs remain a concern because pasture degradation has been observed when fertilization is reduced or eliminated. For this reason, they have identified a need for productive, non-bloating legume species in the perennial pastures. The fertility issue has led to an awareness that assessment of the soil biota is important in evaluating the sustainability of pastures. The authors also point out that management of rehabilitated lands is critical because proper management is important to sustainability. They have also determined that if trees are desired in the rehabilitated landscape then soils of greater depth and enhanced water-holding capacity will be necessary.

These three methods of assessing rehabilitation success (regulatory, utilitarian and ecosystem) show different levels of development. It is important to note that in all three cases there are elements of the utilitarian approach to assessing success. It is also clear in the regulatory and ecological approaches that 'reference or analogue' areas might be used as a guide but one should be careful not to strive for the same 'level or value' for the rehabilitated site because of the age differential between the two systems. The science of assessing rehabilitation success is still being developed and will require much further testing and modification. It is obvious from all three approaches that long-term sustainability is the ultimate goal. With this in mind, much more attention should be given to the dynamics of nutrient cycling to ensure that these systems are sustainable. In countries where land rehabilitation has only recently been considered necessary following disturbance, landscape/resource stability has been the main goal and criteria for assessing rehabilitation have not been developed. At a recent land rehabilitation conference in the People's Republic of China several Chinese speakers made reference to the need to reclaim disturbed and degraded lands to agronomic productivity because of the limited agricultural land base and the large population. In such cases, then, successful rehabilitation will be measured in terms of food production. *Rehabilitation is a very important part of the whole process of mining, but there can only be practical realization of the need to assess that rehabilitation where and when there is economic stability and environmental issues have received the attention of the general population.*

Conclusions

The science and practice of mined land rehabilitation has advanced significantly over the past three decades. The level of understanding and acceptance of this technology varies between countries depending on the age of the science and the political/social conditions. Economics plays a significant role in deciding whether rehabilitation of mined lands is a common requirement or an unaffordable luxury. In countries where basic human needs such as food, housing and health are day-to-day concerns, land rehabilitation and even other environmental concerns are not given high priority. However, it is becoming ever more apparent that the world is recognizing the importance of land rehabilitation from an environmental sustainability standpoint. The demand for food, fibre, good quality water and breathable air is becoming more important to all peoples as the world population expands. Mined land rehabilitation technology is probably at a maturing stage and is being applied in the developed countries but to only a limited degree in developing countries. There is a good knowledge base for addressing the issues related to the conservation of topsoil, toxicity in mine spoils, and the stabilization of the land resource through landscape design and vegetation establishment. However, gaps still exist in understanding how to ensure long-term sustainability of the rehabilitated plant communities and re-establish biodiversity. In particular, the dynamics of below-ground processes in rehabilitated lands is poorly understood, except in regard to arbuscular mycorrhizae. In the future, more emphasis should be given to understanding nutrient cycling on rehabilitation mined surfaces and developing practices necessary to ensure these systems are self-renewing.

Soil compaction is probably still the single most frequent cause of rehabilitation failure because it leads to excessive runoff and restricted plant growth. Compaction is particularly critical for mined croplands. Therefore, greater attention must be paid to examining the relationships between mining operations, land rehabilitation and soil compaction.

Countries that have developed successful land rehabilitation technologies must be aggressive in transferring that technology to less developed countries. Developing countries do not have the financial means to initiate rehabilitation research programmes. Technology must be adapted and adopted from climatically and edaphically similar regions. Rehabilitation of mined lands throughout the world will not only restore production to these lands but also reduce or eliminate major environmental problems that are quickly becoming global concerns.

Public policies developed to assess and ascertain rehabilitation success must be realistic, easy to apply and must not rely on reference or analogue sites. Rehabilitation success criteria must consider ecological factors and should include assessing ecological trends and use utilitarian approaches. The ultimate goal of mined land rehabilitation is to develop a desired plant community or crop production system that is sustainable and meets the intended post-mine land use or need.

References

Bellairs, S. and Davidson, P. (1999) Native plant establishment after mining in Australian rangelands. In: Eldridge, D. and Freudenberger, D. (eds) *People of the Rangelands. Building the Future. Proceedings of the VI International Rangeland*

Congress. VI International Rangeland Congress Inc., Townsville, Australia, pp. 962–968.

Giurgevich, B. (1999) Reclamation success standards for coal surface mine lands in the western United States. In: Eldridge, D. and Freudenberger, D. (eds) *People of the Rangelands. Building the Future. Proceedings of the VI International Rangeland Congress*. VI International Rangeland Congress Inc., Townsville, Australia, pp. 954–957.

Redente, E.F. and Keammerer, W.R. (1999) Use of native plants for mined land reclamation. In: Eldridge, D. and Freudenberger, D. (eds) *People of the Rangelands. Building the Future. Proceedings of the VI International Rangeland Congress*. VI International Rangeland Congress Inc., Townsville, Australia, pp. 957–961.

Rethman, N.F.G., Tanner, P.D., Aken, M.E. and Garner, R. (1999) Assessing reclamation success of mined surfaces in South Africa with particular reference to strip coal mines in grassland areas. In: Eldridge, D. and Freudenberger, D. (eds) *People of the Rangelands. Building the Future. Proceedings of the VI International Rangeland Congress*. VI International Rangeland Congress Inc., Townsville, Australia, pp. 949–953.

Tongway, D.J. and Murphy, D. (1999) Principles for designed landscapes and monitoring of ecosystem development in rangelands affected by mining. In: Eldridge, D. and Freudenberger, D. (eds) *People of the Rangelands. Building the Future. Proceedings of the VI International Rangeland Congress*. VI International Rangeland Congress Inc., Townsville, Australia, pp. 945–948.

Vicklund, L.E. (1999) Domestic livestock grazing as a reclamation tool. In: Eldridge, D. and Freudenberger, D. (eds) *People of the Rangelands. Building the Future. Proceedings of the VI International Rangeland Congress*. VI International Rangeland Congress Inc., Townsville, Australia, pp. 968–971.

Williams, R.J. and Lane, A.M. (1999) Wet season burning as a fuel management tool in wet–dry tropical savannas: applications at Ranger Mine, Northern Territory, Australia. In: Eldridge, D. and Freudenberger, D. (eds) *People of the Rangelands. Building the Future. Proceedings of the VI International Rangeland Congress*. VI International Rangeland Congress Inc., Townsville, Australia, pp. 972–977.

Accounting for Rangeland Resources

12

Paul E. Novelly and E. Lamar Smith

Introduction

Accounting for rangelands implies monitoring, and monitoring is the means for detecting change in resource accounts (Jordaan *et al.*, 1999). While there has been considerable debate about specific monitoring techniques and the theoretical basis for interpreting the data, there has been less focus on assessment of change in rangeland resources across various spatial scales. In particular, it is necessary to examine whether data acquired at one scale can or should be aggregated for use at a higher scale.

There are three important scales relevant to rangeland resource accounting: the 'paddock/property' scale (the enterprise level), *the 'landscape' scale* (focusing on ecological function), *and the 'regional/national' scale.* Each of these scales warrants individual consideration. However, issues and data needs also cross scales. The connectivity between scales is an important but neglected consideration.

This chapter presents the current interpretations of 'scale' for rangeland accounting and the implications for monitoring associated with the chosen scale.

Definition of the Scales

The definition of scale is complicated. The three scales cannot be adequately defined in terms of map scale, i.e. large, intermediate and small, nor are they always distinct.

The 'paddock/property' scale is the ranch (commercial), park (conservation) or other local-level unit managed for a specific objective. The unit size and management objectives can vary. *At this scale the main purpose is to monitor management impact with respect to the desired outputs and, through adaptive learning, to determine if management should change.* At this scale tactical decision-making is the major part of management and can have a direct influence on the profitability of an enterprise.

At the 'regional' scale, assessment characterizes the average and range of conditions over a larger (often administrative) area which encompasses many management units with different ownership and varying management objectives. *At this scale, information is used to guide the framing of resource policy and operational budgets for a nation or subdivision thereof.* Data may be used by whole industries or regional administration to address the environmental and economic implications of change in resource use (Shaver *et al.*, 1999). Decisions taken have a wide impact across many enterprises, and so may affect some enterprises or groups within the region more than others.

The intermediate 'landscape' scale (West, 1999) *differs in concept from the others, being based on ecological function.* Here it is not so much a question of 'size', but of dealing with specific patterns and ecological processes (Ludwig *et al.*, 1997) rather than average values of condition or trend. Interest may extend across property boundaries, but not usually to the regional level. Issues such as biodiversity, habitat fragmentation, species conservation, and catchment or watershed condition may require information across a mixture of land types and ownerships.

Data Needs and Concerns

There are two types of data needs. The first covers '*sustainability*', the basic long-term capacity of the resource to produce goods and services. *Loss of 'productive' capacity stems from a decline in soil quality through erosion, altered infiltration characteristics and nutrient loss* (Tugel, 1999). Therefore, assessment generally involves consideration of rangeland 'health' or condition. Since productive capacity is site-dependent (soil, landform, climate), stratification by land type is required. Loss of ecological services involves many factors but a major one is extinction of species or populations.

The second type of data covers *management effectiveness*. At the property scale, it is important to know whether management decisions are achieving the desired outputs. Again, one of the desired goals is, or should be, protection of productive capacity. However, the main goals of managers are generally increased forage for livestock, improved habitat for wildlife, water quality and so on. Biological diversity is less likely to be the concern of the individual property owner because it has no obvious direct relationship to management objectives, and usually cannot be evaluated without considering a broader geographic area. Unfortunately, such data are 'often collected by individuals and thus can suffer the vagaries of inconsistency, infrequency, peculiar techniques and poor data management' (Jordaan *et al.*, 1999).

Sustainability and management are also paramount for regional/national assessment. Here, the focus is on providing information for setting government policy and budgets, and informing the community about change in the productive capacity of natural resources. Whether government owns the land, regulates its use, or provides incentives for private landowners to implement conservation practices, it needs information to determine the effectiveness of its programmes and overall management. The balance between conservation goals and production goals at a national scale will depend on the amount of central planning carried out in the country. In any case, national level interest is mainly on the general status and variability of condition or trend over large classes of land types or ownership categories, although the 'range in condition' is also sometimes important. Collecting information on individual properties is assumed to be done by other jurisdictions for different purposes.

Although difficult, general assessment of sustainability must be separated from the assessment of rangeland for a specific resource use or uses, or for the production of specific outputs. However, these two issues are often confused. For example, assessment of 'rangeland health' or 'proper functioning condition' of rivers and streams may also assess suitability of the area for wildlife habitat or focus on the presence of 'exotic' plant species. While these factors are important in overall ecosystem sustainability, their relationship with range condition is poorly understood. Instead they reflect the broader management objectives or values held by certain groups. Failure to separate sustainability and conservation goals (which are legitimate because they are presumably in the public interest) from the management goals held by different interest groups (e.g. graziers, hunters, conservation biologists, etc.) can distort information and lead to poor policy decisions.

Public vs. Private Viewpoint

The capability and willingness of either landowners or lessees of public land are uncertain. They have two main motives for monitoring: (i) to guide management; and/or (ii) to prove to government agencies, environmental interests and/or other regulators that their management is sustainable and compatible with public goals.

There is sometimes an attitude that monitoring, even if done by managers, is more about assessing resource management than about enterprise management. Moreover, in some societies, land managers often attribute natural resource degradation to fate (Meyer, 1998). Such views will change, with managers linking management decisions to change in resource condition. While consensus exists that range condition, productivity and livestock weight gains are related, they are rarely linked by monitoring systems and often excluded from business procedures. Although short-term gain can sometimes be maximized by non-sustainable management, good range management is synonymous with good business management and economic survival in the medium and long term. *Land monitoring should now become an essential part of the business decision-making process to optimize long-term financial and land sustainability.* Indications provided by monitoring may be adverse to *short-term* financial return, but positive for *longer-term* success.

Property-level monitoring will document livestock utilization levels, plant composition, ground cover or other attributes related to range condition. However, it is not always apparent to the manager that such measurements translate into livestock carrying capacity, off-take rates, death loss, etc. Consequently, such information is considered 'nice to know' but not particularly relevant to the economic operation. Range scientists must establish the link between resource condition and livestock or wildlife outputs, and make these known to the land manager. Moreover, the imperative to monitor must be based on the land manager's values, not someone else's. Once managers are convinced that monitoring data are really useful, and they have been introduced to monitoring in their own 'language' they are generally willing and able to do their own monitoring.

Experience with public land managers suggests many are willing, even anxious, to do monitoring themselves, or hire someone to do it for them, when they feel threatened by regulatory agencies or environmental organizations. The utility of such information for the livestock business is immaterial. Rather, it is seen as *insurance against the prospect of losing or compromising the viability of their grazing business.* Thus, there is an economic incentive to conduct monitoring related to the requirements or issues likely to be at stake. However, interest wanes when the issues change in what are perceived to be arbitrary ways. For example, a public land manager may invest considerable time or money monitoring plant composition or ground cover (variables important to the regulatory agency), only to be attacked on the alleged effects of management on water quality or a newly listed endangered species.

The question is repeatedly raised as to whether land managers should be expected to conduct their own monitoring to provide information desired by the 'public', when such data are not seen as useful by the land manager. Managers may need to be compensated for providing this information. Some land managers are also concerned that this information may be used to compromise private property rights or lease tenure. This is a complex issue and solutions will depend on the legal basis of land ownership and lease tenure in particular countries, and the purpose for which the information is used.

Regional assessments are used to set policy and/or government programmes, and do not generally relate to individual property management. Such assessments are generally done by government agencies and should respect the rights to privacy and property of landowners. If government agencies want managers on either private or public land to provide such information to them, compensation for added effort may be appropriate.

If the information gathered relates only to the effectiveness of management in meeting conservation (duty of care) and production goals, the situation is different. It is the responsibility of both the private landowner and lessees of public land to obtain such information, although in many countries technical assistance from the government is offered in this process. No compensation for collecting such data should be expected since it is for the exclusive use and benefit of the enterprise.

Where land is leased from the government, lessees must care for the public land. Two things are needed:

- The lessee's acceptance that he/she manages public land and is responsible for its status.
- Agency and public acceptance that monitoring demands placed on lessees must be consistent, and cannot be open-ended.

Therefore, *while some public land lessees may not consider monitoring to be of benefit to their grazing business, the requirement to demonstrate how well public land is being managed is not something for which full compensation to the lessee is appropriate.*

Where grazing businesses are composed of a mixture of private property and government land leases, the situation is more complex. There may be conflict between the public's right to know how public land is managed and the individual's rights to privacy on private property (although a duty of care still exists on the latter).

Aggregation and Disaggregation of Data

Property-level monitoring, generally aimed at obtaining tactical advice, has been practised and promoted for many years, and is usually conducted at representative locations (key areas). Such locations purport to represent characteristic land types, and are expected to demonstrate effects typical of the management applied. Attributes selected for measurement, the season or frequency of measurement, and the way measurements are interpreted depend on the vegetation community being measured, management objectives and time available. Therefore, the type of monitoring conducted varies among properties and management objectives. For example, the managers of grazing businesses and national parks would not necessarily collect the same kinds of data, even when the same land types are involved. However, the attributes of landscape function will be common to both needs.

Interest in regional/national assessments has increased in line with growing 'environmental' concern by the public and the demand to monitor natural resource 'health' nationwide, or even worldwide. Two approaches have been used. The first is to *'scale up' property-level data* to reflect conditions over a much larger area. For example, in the USA, the Bureau of Land Management and the US Forest Service (government agencies managing federal grazing lands) have reported on the general 'condition and trend' of federal rangelands by aggregating information collected on individual grazing allotments. Agencies have used these to document progress, or lack of progress, towards improved range management. These reports have been used to justify the agencies' needs for more resources and to justify demands for more restrictive regulation of grazing. In reality, however, such reports have generally not provided valid or useful information. In part, this results from differences in methodology among or within agencies and over time, making valid comparisons in time or ownership impossible. Another problem is that monitoring data are not representative of all federal grazing lands because monitoring efforts have been concentrated on the areas where problems are known or suspected to exist. Consequently, unused or lightly grazed areas and well-managed allotments are often under-represented.

The second approach is to *monitor large areas separately from the property level using different techniques* (e.g. remote sensing). In the USA, the Natural Resources Conservation Service (NRCS) – a national agency providing technical assistance to private landowners – has conducted the National Resource Inventory (NRI) on private lands for several years. This is a statistically designed nationwide sample to document conditions and required land treatments as a guide to NRCS policy and programmes. Recently, partly at the recommendation of the

Society for Range Management and the National Research Council, there was an effort to expand this type of survey to federal rangelands (Shaver *et al.*, 1999). This statistical sampling approach does not provide data relevant to management of individual properties; that is, it cannot be disaggregated to the property level, but does provide a valid indication of conditions at a regional/national level, or by major ecological zones or categories of land ownership.

Since data collection protocols are, or should be, specific to the goals for which they are collected, aggregation of property-level data to higher levels, and vice versa, is either not feasible or will not provide consistent, interpretable results, unless specifically planned for in terms of the number of samples obtained. Scales of interest are distinct and should be recognized as such. *Efforts to create 'report cards' at the regional, national, or international level should be based on statistically sound approaches designed for each specific purpose.* Agencies responsible for such reports must accept restrictions posed by the data type and not attempt inappropriate interpretation, while advice generated from such data must be qualified by its limitations. Inappropriate use of monitoring data is rife, often presenting either over-optimistic or over-pessimistic views. Monitoring data should primarily be used at the scale at which it was determined, and extreme caution should be exercised when applying it to other scales.

Confusion at the 'Landscape' Level

Rangeland assessment at the landscape level is relatively new. Interest in developing concepts and techniques at this level has arisen largely because of a better understanding of the scale at which rangelands function and, at that level, concerns over biological diversity, habitat conservation for threatened or endangered species, and efforts to implement 'ecosystem management'. This allows consideration of entire catchments or other ecological zones crossing property and jurisdictional boundaries. The field of 'landscape ecology' has developed in response to these issues, and remote sensing and computer technology (e.g. GIS) make it possible.

However, landscape-level attention is not new. For example, land system surveys in Australia since the 1940s have mapped and characterized landscape units based on recurring patterns of land types (for example Stewart *et al.*, 1970). These units are still useful for landscape-level issues (Smith and Novelly, 1997). Classification of wildlife habitat has also considered the pattern of occurrence of vegetation providing for various requirements of wildlife species, for example cover, food, water, etc. However, in both examples, 'landscape'-level assessment was based on consideration of the pattern of discrete land or vegetation types. Some of the newer technologies (West, 1999) involve 'synoptic' measures only obtainable with remote sensing technology, and which are a function of the scale and resolution of the technology applied.

Both property-level and regional/national-level assessments are usually based on point sampling, i.e. data collected on relatively small, site-specific locations, extrapolated to larger areas based on statistical inference or professional judgement. Both are based on similar data and differ mainly in sampling approach and intensity. The landscape level can be considered not merely an intermediate scale between property level and regional, but a conceptually different kind of information unavailable from point sampling, with emphasis on patterns and/or synoptic measures. Landscape type data may be used for regional/national reporting and policy formulations, or as a basis for implementing and monitoring management on large properties or administrative units (e.g. national forests), catchments, or planning areas.

Because of the limited research done on landscape-scale assessment, much remains to be learned. In particular, information linking the synoptic measurements possible with remote sensing technology to actual ecological processes and resource outputs is vital. In other words, while many 'measurements' are possible, there is often insufficient information to select the most useful measurement due to a lack of knowledge of what the measurements mean. Without such information, personal bias and values tend to come into play and for decisions to be based on processes different from the accepted norms of scientific method and/or ethics (West, 1999).

Need to Keep Values and Data Separate

Scientists endeavour to make unbiased observations and measurements. However, there is concern among range scientists about the use of 'junk science' to justify rangeland policy and management direction. Moreover, there is concern about the merging of values and science, which some see as desirable and inevitable, and others view as compromising the usefulness of science (West, 1999).

Value-oriented science may be 'junk science' if the 'scientist' deliberately falsifies data or misrepresents findings to support a particular belief. However, it is inevitable that personal values will influence the conduct of research, even when the scientist attempts to be ethical and objective. There is no intrinsic 'bad' or 'good' in ecological systems, and personal judgements inevitably occur. Personal values affect the questions asked and the data considered relevant to answering them. This extends to rangeland monitoring as well as research, particularly given that 'range condition' is a subjective concept, defined by assembling biophysical variables and making a value-laden interpretation of their meaning. But who makes the choices? 'Traditional' range managers focus on monitoring attributes they consider relevant and interpreting the results against their own paradigms, while others (e.g. conservation biologists) may focus on entirely different attributes and/or interpret data in different ways. This is probably unavoidable. In 1999, Heitschmidt and Klement, (Heitschmidt, personal communication) examined three review papers on grazing effects. Each reviewer cited over 120 references, but only six were cited by all three. Heitschmidt concluded that each author had fairly reported the conclusions given in the papers reviewed, but obviously had used considerably different views of what was relevant or useful in selecting those papers.

While it is perhaps too much to expect all scientists or range managers to be totally objective in selecting what, when, how, where, and how often to measure, *the attributes measured must be adequately defined so that others may form their own opinion of the relevance of the attribute to the question being asked.* Measurements should avoid incorporation of value judgements so that data can be re-interpreted by other observers with different values, or as values or the knowledge base change over time (Ludwig *et al.*, 1997). Subjective observations incorporating value judgements do not represent 'data' for monitoring. For example, rating ground cover as 'adequate' or 'inadequate' for soil protection relies entirely on the opinion of the observer and the standard of adequacy chosen. Estimating percentage ground cover gives data that can be re-measured or re-interpreted by different observers provided that 'ground cover' is adequately defined. Likewise, in assessing condition ratings by observing and weighting multiple variables, the views of the observer are incorporated in determining the result, in that he/she determines the relative importance of each variable. If values and biases of individuals and agencies cannot be prevented from entering into the 'science' used to support their objectives, the data used must be defined and measured objectively, so that those who disagree can repeat the studies elsewhere or re-interpret the data at a later time.

The Time Factor

We have focused on differences in spatial scales. However, different temporal scales also exist for various purposes and interpretations. Some measurements are highly subject to season of the year, for example, or vary greatly between years (e.g. biomass, litter cover). The rate of ecological processes involved also affects the frequency with which monitoring should be conducted and the amount of change to be expected. For example, monitoring of annual plants or insect populations may have a very different requirement with respect to season or frequency of measurement than monitoring changes in large herbivore or tree populations.

The temporal scale should be added to the three spatial scales. This is important for seasonal changes, and the impacts of droughts, fires and other episodic events. Complex systems change more slowly than simple ones, and 'accounting' timescales must reflect this. Some plant and soil attributes change quickly, others slowly. Decisions relevant to each scale (e.g. tactical decisions, long-term planning) should be based on monitoring data that are appropriate to that scale. For example, long-term increase in woody species is of no use for setting this year's stocking rates in relation to last year's and so on.

Conclusions

'Accounting for rangelands' implies a subjective interpretation, although one hopefully based on sound professional judgement or understanding. Although there is a long history of efforts to assess and monitor rangeland, approaches are still evolving from both conceptual and practical standpoints. Most early effort was applied to paddock-level assessment to guide management. This need still exists and has grown so that now many land management or extension agencies no longer have the capacity to meet it. It is clear that landowners or lessees must take responsibility for much of this type of accounting, but relevance to their economic interests must be demonstrated.

Increased environmental awareness and concern by the public has led to increased demand for national/regional accounting relative to sustainability and biodiversity. Attempts to answer this need by simply aggregating data collected for property-level management purposes have not worked. Distinct, statistically based approaches are being developed for this purpose, but more effort is needed.

Finally, the necessity to look at certain processes on a landscape, rather than site-specific, scale is increasingly appreciated. Development of GIS and remote sensing technology has made it possible to address processes at different scales. Much work remains to be done to develop this area conceptually and to integrate this information with traditional assessment procedures and management needs.

Protocols must be developed to define how data can be consistently used across the various spatial scales. However, incorporation of more than just biophysical data into the decision-making process has been as disappointing as the use of biophysical data at inappropriate scales.

Any programme developed to provide 'accounting' information for rangelands must, where appropriate, include as broad a range of information as possible. The data for such programmes must also be seen as part of the decision-making process. Most importantly, however, programmes must be defined to answer specific questions, and must not be considered appropriate to answer a generic 'grab-bag' of questions that evolve over time.

References

Jordaan, F.P., Watson, I.W. and Booysen, J. (1999) Monitoring and resource accounting at the paddock and property scale. In: Eldridge, D. and Freudenberger, D. (eds) *People of the Rangelands. Building the Future. Proceedings of the VI International Rangeland Congress.* VI International Rangeland Congress Inc., Townsville, Australia, pp. 721–726.

Ludwig, J.A., Tongway, D.J., Freudenberger, D., Noble, J.C. and Hodgkinson, K.C. (1997) *Landscape Ecology Function and Management: Principles from Australia's Rangelands.* CSIRO, Melbourne.

Meyer, T.C. (1998) Interactions between livestock farming, human needs and the environment in the communal farming sector – perceptions of field workers in the Ganyesa District of the North West Province. *Policy-making for the Sustainable Use of Southern African Communal Rangelands,* 6–9 July 1998, University of Fort Hare, South Africa.

Shaver, P.L., Spaeth, K., Pellant, M. and Mendenhall, A. (1999) National and regional level rangeland assessment. In: Eldridge, D. and Freudenberger, D. (eds) *People of the Rangelands. Building the Future. Proceedings of the VI International Rangeland Congress.* VI International Rangeland Congress Inc., Townsville, Australia, pp. 736–740.

Smith, E.L. and Novelly, P.E. (1997) *The Assessment of Resource Capability in Rangelands.* LWRRDC Occasional Paper Series No. 08/97.

Stewart, G.A., Perry, R.A., Paterson, S.J., Traves, D.M., Slatyer, R.O., Dunn, P.R., Jones, P.J. and Sleeman, J.R. (1970) *Lands of the Ord-Victoria Area Western Australia and the Northern Territory.* CSIRO, Australia Land Research Series No. 28.

Tugel, A.J. (1999) Soil quality institute initiatives and rangeland assessments. In: Eldridge, D. and Freudenberger, D. (eds) *People of the Rangelands. Building the Future. Proceedings of the VI International Rangeland Congress.* VI International Rangeland Congress Inc., Townsville, Australia, pp. 758–759.

West, N.E. (1999) Accounting for rangeland resources over entire landscapes. In: Eldridge, D. and Freudenberger, D. (eds) *People of the Rangelands. Building the Future. Proceedings of the VI International Rangeland Congress.* VI International Rangeland Congress Inc., Townsville, Australia, pp. 726–735.

Building on History, Sending Agents into the Future – Rangeland Modelling, Retrospect and Prospect

13

Timothy J.P. Lynam, Mark Stafford Smith and William J. Parton

People, Theory and the Evolution of Modelling in Rangeland Science

Rangeland science was complex enough when scientists had only to contend with animals and their feed resources. Now that science has put people centre-stage, rangeland scientists have to contend with a world that has become much more complex, much less certain, and spans a far greater range of issues, problems and interests. The globalization of markets and trade opportunities, as well as of policy and environmental issues (such as climate change), has meant that local-scale models or observations are no longer adequate to address natural resource management issues.

For the modeller, including people greatly increases the difficulty of representing the system adequately. There are two aspects of this problem. First, in developing management support tools for managers or policy-makers, the modeller has to understand what information the decision-makers use or require, as well as understanding how they reach their decisions. Second, in analysing how large-scale systems behave, people become a part of the model. This means that the modeller must understand and model not only the ecology, but also the social and economic components of the system.

At the same time, rangeland scientists have lost the reassuring and predictable trajectories of equilibrium theory; they are faced instead with systems characterized by uncertainty and surprise, non-linearities and multiple trajectories with multiple stable states. Equilibrium theory, with its purported universal simplicity, has made possible the rapid advances in rangeland understanding in the 1960s and 1970s, but proved inadequate and has lost much of its influence as more realistic views of non-equilibrial systems have become the new dogma. More recently, even the comfort of multiple stable states is being replaced by views of complex adaptive systems – systems that evolve, in which even the concept of equilibria has little meaning (Allen, 1992; Janssen, 1998).

Disciplines such as rangeland science go through phases in their evolution. Working within the Cartesian reductionist paradigm of breaking complex problems into simpler constituent parts to aid investigation and understanding, disciplines begin with a powerful *analytical* phase. Some of the most remarkable achievements of contemporary science have emerged from this phase. But this phase exacerbates the distancing of research from real world issues, leading to demands for *synthesis* – the need to put the reductionist understanding back together into some sense of coherence that is better equipped to deal with real world problems and issues.

In this chapter, we examine how rangeland modelling has emerged from its roots in reductionist excellence to stand on the threshold of a new 'agent-based' modelling phase. The threshold offers daunting challenges and exciting opportunities. We start this process by examining rangeland modelling in its *analytical* phase, past and present. Thereafter, we explore where we believe rangeland modelling is headed as it grapples with *synthesis*. We conclude with some of the challenges and opportunities that face the discipline in this synthetic phase.

Rangeland Modelling in the Analysis Phase – Unpacking the Boxes of Rangeland Systems

Our objective is not to review rangeland systems modelling in detail (e.g. see Hanson *et al.*, 1985, or Ågren *et al.*, 1991), rather to provide a general sense of the trends in rangeland modelling and to relate these to the analytical phase of rangeland science. There are many branches of rangeland modelling that we will not cover, such as the often-used statistical model of Coe *et al.* (1976), used to predict animal biomass based on rainfall. We focus instead on the use of process modelling to improve understanding of rangeland structure and functioning, and provide information for decision-makers who manage rangelands at both land management and policy levels.

Two threads of modelling may be defined in relation to rangeland systems. The 'ecosystem thread' concerns modelling efforts directed towards understanding the ecological processes of rangelands. The 'managerial thread' has, as its central concern, the provision of management or policy-related information or solutions. The ecological thread was the main initial focus of rangeland modellers. The classic example was the grasslands systems model ELM developed by Innis and others in the mid-1970s (Innis, 1978). In this model, Innis and colleagues explored the complex interaction of range plant production and species composition with nutrient cycling, soil and temperature dynamics, and animal production and consumption. The resulting model was useful as a research tool but too difficult to use as a practical management tool.

In a similar vein the animal science systems group at Texas A&M University (TAMU) developed systems models of cattle herds in given forage and management situations (Sanders and Cartwright, 1979). These models did not deal with the individual components of ecosystems as much as the ELM model, but the TAMU model still contained considerable detail about the ecological components of the cattle–forage system. Whilst both models were explicitly expected to deal with management issues, however, neither of them had people, human decision-making or finances built into their structure. The same could be said of the equivalent Australian models that were developing around the same time (GRASP – McKeon *et al.*, 1990; GRAZPLAN – Moore *et al.*, 1991), and the herd dynamic model of Konandreas and Anderson (1982) in Africa.

The trend of increasingly detailed investigations of the substance of ecosystems continued. CENTURY (Parton *et al.*, 1987) expressed the then current understanding of plant production, nutrient cycling and soil carbon dynamics in a sophisticated simulation model, which has continued to evolve to this day. FORAGE (Baker *et al.*, 1992) was a model of beef cattle feed intake which was designed to couple the forage production capabilities of a sophisticated pasture production model (SPUR; Hanson *et al.*, 1988) with a derivative of the TAMU cattle model.

These modelling efforts demonstrated the increasingly detailed understanding and modelling that were at the forefront of rangeland science. Although they were complex, they included simplifications derived from even more complex models of some components of the rangeland ecosystem, like ELM, and from other new research results. However, this substantial model simplification was offset by complex representations of additional processes, so that they remain difficult to use as management tools.

There were other initiatives and directions in rangeland modelling. A notable sub-thread of the ecosystems approach to modelling was the spatially explicit SAVANNA model developed by Coughenour (1992).

Based on his extensive experience in East African ecosystems, Coughenour recognized the need to deal with spatial variability in rainfall and landscape characteristics, and hence plant growth and production, in order to predict the productivity of rangeland systems in East African savannas realistically. In order to move from a point model to a spatially explicit representation, Coughenour had to simplify (cf. Hall *et al.*, 1999). He reduced the detail in the animal component of the model – dealing with herds rather than individual animals. He also reduced the consideration of nutrient cycling and soil organic matter dynamics, and focused on the effect of soil water dynamics on rangeland and tree production. SAVANNA is now being used elsewhere (e.g. Ludwig *et al.*, 1999). Noble (1975) created a spatially explicit model of a sheep paddock of similar complexity.

The effort required to create large and complex models of whole systems means that few have been made. Whilst they have proven invaluable as an aid to improving our understanding of the system, the incredible complexity of interactions which can emerge means that they are very hard to validate properly (Parton, 1999) and inefficient to use for other purposes. More recent rangeland modelling efforts (e.g. Eldridge and Freudenberger, 1999) have concentrated on addressing narrow and relatively specific problems.

It is also abundantly clear that it is the simple, readily understood and (in their narrow domain) well-validated models which have the greatest use and support in actual decision-making by managers and policy-makers alike. These constitute models in the 'managerial thread' of model development, which has been proceeding nearly as long as the 'ecosystem thread', with some overlap between the two. A major review workshop in 1991 (Stuth and Lyons, 1993) examined a wide range of models in the managerial thread, as well as the interface between the two. It began to highlight the importance of linking 'soft' and 'hard' decision support systems into a single delivery package. The importance of recognizing the target audience was also discussed at length – it is a rare and serendipitous occasion when models constructed for research readily translate into decision support tools (Stuth and Stafford Smith, 1993). Models built to enhance scientific understanding tend to be complex and require specialist support, whilst models aimed at assisting decision-makers must be transparent and easily used. Some topics and some decision-makers can encompass both, more often in the policy realm than on-farm, but this is the exception rather than the rule. Limited uptake of models in day-to-day rangeland management is largely a result of the failure to recognize this (Campbell, 1999). However, these types of models are likely to take centre-stage as rangeland modelling enters a new synthetic phase.

Rangeland Modelling in the Synthesis Phase –
Standing on the Threshold

Looking into our crystal ball of rangeland modelling in the coming decades we see four major evolutionary lines.

1. The *'mega-systems number crunchers'* are an evolving continuation of the current schools of systems modellers. Perhaps best characterized by SAVANNA (Ludwig *et al.*, 1999), SPUR 2000 (Pierson *et al.*, 1999), GRASP (Timmers *et al.*, 1999), GRAZPLAN (Freer *et al.*, 1997) and HILLPLAN (Milne *et al.*, 1999), this evolutionary line will become characterized by spatially explicit models, often linked to or integrated with a GIS, and will incorporate the major elements of rangeland systems. They will increase in complexity as they incorporate more of the human dimension and are made more user-friendly. Problems of parameterization and steep learning curves will severely limit their use outside of case-study work for policy circles. They will generally be used by competent teams of technicians who can provide all of the support necessary to parameterize and run these models, and then interpret the results. They will continue to be important in helping to improve our understanding of the system, but we do not see this line as being a dominant species in the modelling landscape of the next few decades, even if they can be modularized (Parton, 1999).

2. The *'local knowledge representations'* are those models perhaps best characterized by Starfield and Bleloch's (1991) simple management-oriented models, together with the emerging class of statistical models called Bayesian Belief Networks (BBNs, e.g. Jensen, 1996). These models are characterized by their focus on simple and specific problems or issues, and by their incorporation of managerial or indigenous knowledge in model development and testing. Most often these models are knowledge representations for use by managers to guide their activities (e.g. Collis and Corbett, 1999) or to guide scientists in the design and implementation of research programmes or experiments. We see this evolutionary line as being a key species on the modelling landscape; cross-breeding between the two main sub-species (BBNs and simple simulation models) may create a serious contender for a dominant species!

3. *'Complex adaptive agents'* models are the most glamorous species on the landscape but they may easily become dinosaurs like the bigger systems models if their inherent maladaptations are not overcome. 'Agents' are computerized entities emerging naturally from object-oriented programming techniques, which can interact with one another and exhibit evolving behaviour in the arena of their electronic modelling runs. This evolutionary line is characterized by models such as Bousquet *et al.* (1999) and Janssen *et al.* (2000), as well as the classic work of Holland (1995). Learning, adaptation and behaviour emerging from the interactions of multiple agents yield rich insights into the behaviour of complex systems. So far this class of models has been most often used to inform and explore theory with some limited policy relevance. A major challenge in the use of these models is to make them management relevant. They are unlikely to be used for specific range management problems, but they could have great relevance in policy analysis. These models could also capture the great opportunity provided by the Internet for agent-based modelling (Klusch, 1999; http://www.agent.org) in the broader sense, as discussed further below.

Another sub-species in this class includes those models based on cellular automata. The simple rule structures of these models have been shown to be capable of simulating most types of complex system behaviour (Wolfram, 1986) – from stable limit cycles to chaos. Although not frequently seen as modelling tools in rangeland science to date, these models are simple to build and have the potential to make important contributions to rangeland modelling. Starfield *et al.* (1993) have taken steps in this direction with their frame-based modelling approach. The hybrid between state-and-transition models and cellular automata represented by the frame-based approach could be a powerful tool in the *complex adaptive agent* modelling fold, which could also cross-breed with the *local knowledge representation* class of models.

4.   The '*simple heuristics*' class of models is likely to be, as it has often been in the past, the dominant species on the landscape. Simple, fast, easy to parameterize and use, these models have proven themselves over the decades to be the model of choice for most decision-makers (and, although they might not admit it, scientists as well!). Typified by examples such as Peter Johnston's 'Safe Carrying Capacity' model (Johnston and Garrad, 1999) and Jarman's 'Feeding Styles' model (Jarman, 1974), these general rules of thumb have had enormous impacts on scientists, managers and policy-makers. It is worth noting that these models most reliably emerge as tested simplifications from the extensive use of more complex models (e.g. GRASP in the case of the Safe Carrying Capacity model; SPUR for STEERISK – Hart, 1991) as well as from extensive field research and analysis in the case of Jarman's model. These simple models aim to capture the critical elements of a complex understanding in rules that are very simple to apply (Campbell (1999) provides some handy guidelines for creating such models).

In reality, of course, these classes of models overlap and interplay. For example, software agents could play roles beyond those of modelling directly in the next few decades, as discussed later. Further speciation and cross-breeding will undoubtedly render these forecasts naïve. However, we have alluded to several challenges that the next generation of rangeland models and modellers face, and it is clear that some of these will be important, whatever detailed direction the models take. Recent discussion highlights the issues of scale, uncertainty and, in particular, the uncertainty associated with using different knowledge sources as having the potential to be among the most serious challenges. In the next section we explore these challenges in more detail, and look at the exciting opportunities that new technology and theories present.

Challenges and Opportunities

The responses that emerge to a number of challenges and opportunities will be key determinants of the direction that modelling of rangelands takes.

Scale[1] seems to be the scientific equivalent of the Holy Grail – an oft-changing target that is never fully grasped! Despite many excellent attempts to address the issue of scale in the ecological literature (e.g. Meentemeyer, 1989; Wiens, 1989), it is hard to deal with in models, particularly where multiple scales are concerned. Yet the ever more recognized complexity of resource management problems increasingly requires models to handle multiple scales.

Models (and indeed most ecological insights) are generally developed at a particular scale, although this scale is seldom explicitly stated. Thus, *even when models are 'multi-scaled', they are most often defined at discrete scales.* We have a poor knowledge of how to extrapolate between scales, and many management failures arise from solutions at one scale failing to take account of effects at another scale. Figure 13.1 illustrates this problem with hypothetical sets of parameter values that change as the scale at which observations are made changes. Each function represents the parameterized value of a different type of variable. As the scale of interest changes, some model parameters change in absolute value whilst others do not, leading to changes in the relative significance of different variables. Some variables may even cease to be relevant at all at some scales. As a consequence, model representations defined at two different scales (vertical lines) will need to represent different variables as well as having different parameterizations of the same variable at the different scales.

Holling (1992) suggests that ecological structures or processes are often clumped or 'lumpy' across scales; the emerging evidence supporting this confirms that any implicit assumption of linear changes in

---

[1] We use the term *sensu lato* here, to refer both to the *grain* or smallest unit of resolution, as well as to the *extent* or largest unit of resolution.

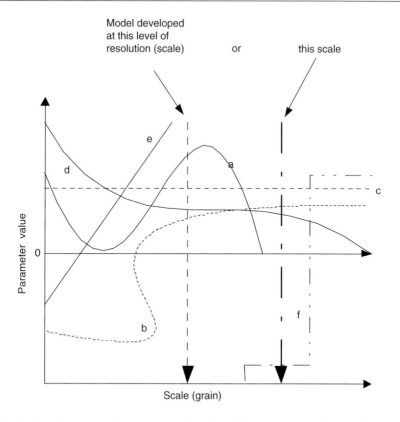

Fig. 13.1. Hypothetical parameter values for different variables (lines a–f) illustrating how these may change with changing scale (grain); models developed at two different scales (vertical lines) may require quite different sets of variables to be represented, as well as different parameterizations of those variables.

parameter values with changing scale is wishful thinking at best. Thus, there are at least two key problems associated with scale in modelling (Allen, 1992): (i) knowing how to interpolate between measured scales when dealing with multi-scale models; and (ii) knowing what variables enter and exit from the model as the scale changes.

Uncertainty is another problem that will require continued attention from modellers of the near future. By no means a new problem in any field of scientific inquiry, uncertainty has a special place in modelling. As George Box (1979) observed: 'All models are wrong but some are useful.' There are many aspects of uncertainty – in model formulation, model inputs, model outputs. Here we focus on two issues – the propagation of errors through a model and how uncertainty is represented in models.

Error propagation has long been a concern in modelling. Many of the current generation of large-scale simulation models are essentially deterministic – given a set of input parameters they generate a given output or solution. This deterministic input–output relationship hides the great deal of uncertainty that is hidden in the structure of the model as well as in the values of parameters and input data values (see for example, O'Neill, 1971). In an investigation of sources of uncertainty in ecological models, O'Neill and Gardener (1978) established that mean parameter values (i.e. those most often used as *the* parameter value in deterministic models) did not necessarily yield the best prediction results. The quality of predictions depended more on the distributions of the parameter values. Thus, implicit in the construction of most models are many assumptions about uncertainty, and about which sources of uncertainty (model structure, model parameter, natural system variability) are most important. These decisions are seldom transparent to the model user, so that the model builder's error assumptions are propagated into the users' beliefs and potentially into actions. This becomes particularly alarming when the model is taken out of the original formulation context, as often occurs when a model originally conceived for the purposes of enhancing understanding is taken into a decision support context.

As models are constructed to include increasingly diverse knowledge sources – *consilience* as E.O. Wilson (1998) has termed it – then the problem of error propagation becomes even greater. Models that incorporate local villagers' beliefs as to likely outcomes, with scientific measures of input levels and model components derived from the literature, encompass very different elements of uncertainty. When these models are developed as complex adaptive models, the problems are compounded and make traditional validation procedures (Naylor and Finger, 1967; Mankin *et al.*, 1975) of questionable usefulness. Users come to rely ever increasingly on the perceived competence of the model builder in the issue domain of concern, and on the degree to which heuristically acceptable results emerge in a gestalt way. Greater attention should be paid to this issue if models are not to fall into another self-imposed credibility gap.

Recognizing uncertainty leads us to another question: *how do we present uncertainty to different audiences in ways that will have meaning to them?* The representation of uncertainty is also not a new problem but it is becoming increasingly important, particularly as scientists interact more directly with managers and policy-makers. Data or information visualization is a large and sophisticated field of endeavour in its own right. As model output is designed to be interpretable by a wider audience (particularly as models are designed to run on the web) then the question of how to present results, and the uncertainty associated with those results, in a meaningful way becomes increasingly problematic.

As rangeland analysts and modellers come to grips with integrating the human dimensions of rangelands into their ecological models and understanding (or vice versa), they will face the severe problem of the limited compatibility, both theoretical and practical, between most current economic models and ecological models. Despite the advances being made by Holling (1986) and others (e.g. Gunderson *et al.*, 1995) in the development of an integrating theoretical framework, this framework does not yet provide specific algorithms for modelling purposes. Most models of economic systems and human decision-making are still based on equilibrium theories or theories of utility maximization. Little progress has been made towards developing models based on more realistic theoretical premises such as those posed by adaptive agents (although the recent developments that link cognitive psychology and economic modelling do hold much promise – see for example, Anderson, 1998 – seeking simple rules that make us smart). This will be an important area of endeavour in the coming decade.

Despite the many challenges that face millennium modellers there are also a number of remarkable opportunities that are a product of developments in both the technology and the conceptualization of models. These opportunities will make the next few decades very exciting for modellers and model users.

Some of the most exciting opportunities are provided by the Internet – the potential for distributed computing, rapid information dissemination and ready access to data sets, open to anyone with a computer and an Internet link. In the simplest sense, this provides better opportunities for interaction. The electronic journal *Conservation Ecology* has already had issues with models that can be downloaded and run from the papers describing them (http://www.consecol.org/). The Century model group has also made use of the Internet for distribution (http://www.nrel.colostate.edu/PROGRAMS/MODELING/CENTURY/ CENTURY.html) and is working on the development of software packages that will make it easier for non-modellers to use the model for applied management problems in rangeland and crop agroecosystems.

However, there are more esoteric possibilities – classes of the previously mentioned agents are now being developed to roam the Internet carrying out tasks for their controllers. In combination with other classes of models, this provides the potential for perhaps the most exciting modelling revolution in decades. Internet agents could be sent on information-seeking missions, perhaps to query a database or to run a web-based version of SAVANNA, by model interface agents who need more information to improve the understanding or learning of agents in a simulation model. The design of intelligent agents is a major area of research in computer science (Weiss, 1999), and agent-based simulations are rapidly becoming the leading edge of modelling.

A further powerful opportunity evolving from these developments is the possibility of linking modellers, policy-makers and managers in controlled simulations, games or decision-making activities. Imagine the scenario where a developing country wishes to revise its policy on land allocations to livestock as opposed to wildlife production systems. A key question in the policy-makers' minds might be how the local villagers would respond to incentives for land reallocation. To find out, one option would be for the policy-makers to distribute a series of inter-active scenario descriptions to village telecentres (rural information centres that contain email, Internet, telephone and library facilities) requesting responses from local leaders. *This type of communication need no longer be a top-down or one-way process.* It could be developed to enable villagers to pose alternative scenarios to policy-makers, with the interaction facilitated through the use of carefully designed, interactive models. Each node in the model (i.e. a telecentre or decision-maker) might be represented by an agent in the model. Each node would programme its agent to respond to the scenarios in specific ways. The villager agents would interact with the policy node so the policy-maker could test different responses to different decisions, but the villagers could also evolve their responses as the policy scenarios change, all without having to commit to major changes on the ground until there was general agreement.

Thus *the Internet and remarkable advances in computing power and technology provide great opportunities that modellers at the start of the third millennium are fortunate to have available to them.* With these opportunities come challenges such as how best to represent uncertainty in order to enable the user to distinguish degrees of truth in the modelling tools they may wish to access. Facing this greatest revolution of recent times, what steps are we as rangeland scientists taking to ensure that we are prepared to make best use of these oppor-tunities? Little strategic thinking along these lines has emerged as yet, although the possibility of improving the formal assessment of new models through Internet access to standard data sets is recognized, as is the need for better testing standards (Parton, 1999). However, just as information brokers are emerging to mediate between naïve users and the vast bulk of unsorted data already available on the Internet, similar credible expert services may be needed to facilitate the use of agent-based models on the web. A future rangeland conference could do well to tackle head-on the question of *how best to capitalize on these developments and whether the rangeland profession is ready to provide such expert facilitation in the service of better management of rangelands.*

Summary and Conclusions

Modelling of rangeland systems has entered a new era in which the problems that modellers attempt to solve, or develop models for, have become increasingly complex. The highly productive period of reductionist investigation has yielded significant advances in our understanding of rangelands and rangeland ecology. As we move into a phase of synthesis, we see four broad species of model populating the rangeland modelling landscape over the next few decades. Some of these will continue to target improved understanding, while others focus more on delivery to decision-makers, whether these are policy-makers or resource managers.

All, however, must handle the human dimension of rangelands better. Modellers face a number of challenges in their attempts to either include humans as actors in their models or develop and present models to managers, policy-makers or lay people. As models encompass more and more of the reality of rangeland systems, modellers must develop solutions to the issue of scale, to dealing with and presenting uncertainty, and to building models of *consilience* – using the knowledge of different sectors of society, not just of scientists.

Indeed, rangeland modelling stands on the threshold of exceptionally interesting times. The opportunities that are presented by the Internet for distributed modelling, for agent-based modelling and for interactive modelling are truly remarkable whilst at the same time presenting considerable challenges to modellers. The technology is there. Are we ready?

Acknowledgements

This chapter draws on the contributions and discussion of participants at the 'Modelling for Better Rangelands' session of the VIth International Rangelands Congress in July 1999; we thank those participants for their valuable insights – the chapter's idiosyncrasies are our own.

References

Ågren, G.I., McMurtrie, R.E., Parton, W.J., Pastor, J. and Shugart, H.H. (1991) State-of-the-art of models of production decomposition linkages in conifer and grassland ecosystems. *Ecological Applications* 1, 118–138.

Allen, P.M. (1992) Modelling evolution and creativity in complex systems. *World Futures* 34, 105–123.

Anderson, J.L. (1998) Embracing uncertainty: the interface of Bayesian statistics and cognitive psychology. *Conservation Ecology* [online] http://www.consecol. org/vol2/iss1/art2 2, 2.

Baker, B.B., Bourdon, R.M. and Hanson, J.D. (1992) FORAGE: a model of forage intake in beef cattle. *Ecological Modelling* 60, 257–279.

Bousquet, F., d'Aquino, P., Rouchier, J., Requier-Desjardins, M., Bah, A., Canal, R. and Le Page, C. (1999) Rangeland herd and herder mobility in dry intertropical zones: multi-agent systems and adaptation. In: Eldridge, D. and Freudenberger, D. (eds) *People of the Rangelands. Building the Future. Proceedings of the VI International Rangeland Congress*. VI International Rangeland Congress Inc., Townsville, Australia, pp. 831–836.

Box, G.E.P. (1979) *Robustness in the Strategy of Scientific Model Building*. Academic Press, New York.

Campbell, S. (1999) The pocket pedagogue: application and development of models for research and management purposes incorporating learning theory. In: Eldridge, D. and Freudenberger, D. (eds) *People of the Rangelands. Building the Future. Proceedings of the VI International Rangeland Congress*. VI International Rangeland Congress Inc., Townsville, Australia, pp. 819–824.

Coe, M.J., Cumming, D.H. and Phillipson, J. (1976) Biomass and production of large African herbivores in relation to rainfall and primary production. *Oecologia (Berlin)* 22, 341–354.

Collis, S.N. and Corbett, J.D. (1999) A dynamic natural resource spatial information system for east Africa. In: Eldridge, D. and Freudenberger, D. (eds) *People of the Rangelands. Building the Future. Proceedings of the VI International Rangeland Congress*. VI International Rangeland Congress Inc., Townsville, Australia, pp. 879–880.

Coughenour, M.B. (1992) Spatial modeling and landscape characterization of an African pastoral ecosystem: a prototype model and its potential use for monitoring drought. In: McKenzie, D.H., Hyatt, D.H. and McDonald, J. (eds) *Ecological Indicators*. Elsevier, New York.

Eldridge, D. and Freudenberger, D. (eds) (1999) *People of the Rangelands. Building the Future. Proceedings of the VI International Rangeland Congress*, Townsville, July 1999.

Freer, M., Moore, A.D. and Donnelly, J.R. (1997) GRAZPLAN: decision support systems for Australian grazing enterprises. *Agricultural Systems* 46, 77–126.

Gunderson, L.H., Holling, C.S. and Light, S. (1995) *Barriers and Bridges to the Renewal of Ecosystems and Institutions*. Columbia University Press, New York.

Hall, W.B., Bean, J., Beeston, G., Dyer, R., Flavel, R., Richards, R., Tynan, R. and Watson, I. (1999) Aussie GRASS: Australian Grassland and Rangeland Assessment by Spatial Simulation. In: Eldridge, D. and Freudenberger, D. (eds) *People of the Rangelands. Building the Future. Proceedings of the VI International Rangeland Congress*. VI International Rangeland Congress Inc., Townsville, Australia, pp. 854–855.

Hanson, J.D., Parton, W.J. and Innis, G.S. (1985) Plant growth and production of grassland ecosystem: a comparison of modelling approaches. *Ecological Modelling* 29, 131–144.

Hanson, J.D., Skiles, J.W. and Parton, W.J. (1988) A multi-species model for rangeland plant communities. *Ecological Modelling* 44, 89–123.

Hart, R.H. (1991) Managing range cattle for risk – the STEERISK spreadsheet. *Journal of Range Management* 44, 227–231.

Holland, J.H. (1995) *Hidden Order. How Adaptation Builds Complexity.* Helix Books, Reading, Massachusetts.

Holling, C.S. (1986) Resilience of ecosystems: local surprise and global change. In: Clark, W.C. and Munn, R.E. (eds) *Sustainable Development of the Biosphere.* Cambridge University Press, Cambridge, pp. 292–317.

Holling, C.S. (1992) Cross-scale morphology, geometry and dynamics of ecosystems. *Ecological Monographs* 62, 447–502.

Innis, G.S. (1978) *Grassland Simulation Model.* Springer-Verlag, New York.

Janssen, M. (1998) *Modelling Global Change. The Art of Integrated Assessment Modelling.* Edward Elgar, Cheltenham, UK.

Janssen, M.A., Walker, B.H., Langridge, J. and Abel, N. (2000) An adaptive agent model for analysing co-evolution of management and policies in a complex rangeland system. *Ecological Modelling* 131, 249–268.

Jarman, P.J. (1974) The social organization of antelope in relation to their ecology. *Behaviour* 48, 215–269.

Jensen, F.V. (1996) *An Introduction to Bayesian Networks.* UCL Press, London.

Johnston, P.W. and Garrad, S.W. (1999) Estimating livestock carrying capacities in south-west Queensland, Australia. In: Eldridge, D. and Freudenberger, D. (eds) *People of the Rangelands. Building the Future. Proceedings of the VI International Rangeland Congress.* VI International Rangeland Congress Inc., Townsville, Australia, pp. 861–862.

Klusch, M. (1999) *Intelligent Information Agents.* Springer, New York.

Konandreas, P.A. and Anderson, F.M. (1982) *Cattle Herd Dynamics: an Integer and Stochastic Model Evaluating Production Alternatives.* International Livestock Centre for Africa, Addis Ababa, Ethiopia.

Ludwig, J.A., Coughenour, M.B. and Dyer, R. (1999) Modelling the impacts of fire and grazing on the productivity and sustainability of tropical savannas, northern Australia. In: Eldridge, D. and Freudenberger, D. (eds) *People of the Rangelands. Building the Future. Proceedings of the VI International Rangeland Congress.* VI International Rangeland Congress Inc., Townsville, Australia, pp. 845–847.

Mankin, J.B., O'Neill, R.V., Shugart, H.H. and Rust, B.W. (1975) The importance of validation in ecosystem analysis. In: Innis, G.S. (ed.) *New Directions in the Analysis of Ecological Systems.* Part 1. The Society for Computer Simulation, La Jolla, California, pp. 63–71.

McKeon, G.M., Day, K.A., Howden, S.M., Mott, J.J., Orr, D.M., Scattini, W.J. and Weston, E.J. (1990) Northern Australian savannas: management for pastoral production. *Journal of Biogeography* 17, 355–372.

Meentemeyer, V. (1989) Geographical perspectives of space, time, and scale. *Landscape Ecology* 3(3/4), 163–173.

Milne, J.A., Sibbald, A.R., Farnsworth, K.D. and Birch, C.P.D. (1999) A model for predicting the impact of grazing animals on animal production and vegetation changes in temperate grasslands and rangelands. In: Eldridge, D. and Freudenberger, D. (eds) *People of the Rangelands. Building the Future. Proceedings of the VI International Rangeland Congress.* VI International Rangeland Congress Inc., Townsville, Australia, pp. 872–873.

Moore, A.D., Donnelly, J.R. and Freer, M. (1991) GrazPlan: an Australian DSS for enterprises based on grazed pastures. In: Stuth, J.W. and Lyons, B.G. (eds)

Proceedings of International Conference on Decision Support Systems for Resource Management. College Station, Texas, April 1991, pp. 23–26.

Naylor, T.H. and Finger, J.M. (1967) Verification of computer simulation models. *Management Science* 14(2), B-92–B-101.

Noble, I.R. (1975) Computer simulations of sheep grazing in the arid zone. Unpublished PhD thesis, University of Adelaide, Australia, 308 pp.

O'Neill, R.V. (1971) *Error Analysis of Ecological Models.* US Atomic Energy Commission, Oak Ridge, Tennessee.

O'Neill, R.V. and Gardener, R.H. (1978) Sources of uncertainty in ecological models. In: Zeigler, B.P., Elzas, M.S., Klir, G.L. and Oren, T.I. (eds) *Methodology in Systems Modelling and Simulation.* North-Holland Publishing Company, Amsterdam, the Netherlands, pp. 447–463.

Parton, W.J. (1999) Use and testing of grassland models: past, present and future. In: Eldridge, D. and Freudenberger, D. (eds) *People of the Rangelands. Building the Future. Proceedings of the VI International Rangeland Congress.* VI International Rangeland Congress Inc., Townsville, Australia, pp. 836–841.

Parton, W.J., Cole, C.V. and Ojima, D.S. (1987) Analysis of factors controlling soil organic levels in the Great Plains. *Soil Science Society of America Journal* 51, 1173–1179.

Pierson, F.B., Spaeth, K.E. and Carlson, D.H. (1999) SPUR 2000: decision support system for assessing and managing rangelands. In: Eldridge, D. and Freudenberger, D. (eds) *People of the Rangelands. Building the Future. Proceedings of the VI International Rangeland Congress.* VI International Rangeland Congress Inc., Townsville, Australia, pp. 840–841.

Sanders, J.O. and Cartwright, T.C. (1979) A general cattle production systems model. I: Structure of the model. *Agricultural Systems* 4(3), 217–227.

Starfield, A.M. and Bleloch, A.L. (1991) *Building Models for Conservation and Wildlife Management.* Burgess International Group, Edina, Minnesota.

Starfield, A.M., Cumming, D.H.M., Taylor, R.D. and Quadling, M.S. (1993) A frame-based paradigm for dynamic ecosystem models. *AI Applications* 7, 1–13.

Stuth, J.W. and Lyons, B. (eds) (1993) *Emerging Issues for Decision Support Systems for Management of Grazing Lands.* MAB-UNESCO, Paris.

Stuth, J.W. and Stafford Smith, D.M. (1993) Decision support for grazing lands: an overview. In: Stuth, J.W. and Lyons, B. (eds) *Emerging Issues for Decision Support Systems for Management of Grazing Lands,* MAB-UNESCO, Paris, pp. 1–35.

Timmers, P.K., Clewett, J.F., Day, K.A., McKeon, G.M., Pinnington, G.K. and Scanlan, J.C. (1999) WinGRASP, a modelling package for pasture growth and grazing systems in northern Australia. In: Eldridge, D. and Freudenberger, D. (eds) *People of the Rangelands. Building the Future. Proceedings of the VI International Rangeland Congress.* VI International Rangeland Congress Inc., Townsville, Australia, p. 881.

Weiss, G. (1999) *Multiagent Systems: a Modern Approach to Distributed Artificial Intelligence.* Massachusetts Institute of Technology, Boston.

Wiens, J.A. (1989) Spatial scaling in ecology. *Functional Ecology* 3, 385–397.

Wilson, E.O. (1998) *Consilience. The Unity of Knowledge.* Abacus, London.

Wolfram, S. (1986) *Theory and Applications of Cellular Automata.* World Scientific, Singapore.

Integrating Management of Land and Water Resources: the Social, Economic and Environmental Consequences of Tree Management in Rangelands

14

Tom Hatton

Introduction

Rangelands are characterized primarily by their aridity, and water to a very great extent defines local productivity and economics, landscape biodiversity and a host of other downstream environmental services. The culture, activities and attitudes of a range of stakeholders living within or downstream of rangeland catchments are intimately associated with the availability of water and vulnerable to changes in water quality and quantity. The potential for conflict among these stakeholders over water issues is high. A well-known phrase from 19th-century American rangelands colourfully sums this up: 'Whiskey is for drinking, water for fighting over.'

While our level of dialogue and analysis regarding rangeland water resources is perhaps more sophisticated and complex than it was 100 years ago, the issues are also more complex, the stakeholders more diverse and the demands on water greater than ever. The hydrological impacts of prior land use, and the new uses of land to which we may put rangelands, have the potential to create new winners and losers (Syme *et al.*, 1999). One such change in land use putting pressure on rangeland water resources around the world is associated with the management or establishment of trees.

©CAB *International* 2002. *Global Rangelands: Progress and Prospects*
(eds A.C. Grice and K.C. Hodgkinson)

The Changing Views on Trees in Rangelands

Over the past decade or so, there has been a dramatic shift in the attitude towards woody vegetation in rangelands, at least in developed nations. For much of the historic period of settlement in the USA, Australia, Mexico and other nations colonized by Europeans, native woody vegetation was widely considered incompatible with farming and grazing systems ('woody weeds') and a great deal of effort was directed towards its clearing and control. Much of the research and development investment in rangeland science was aimed at underpinning this effort. Even in 1999, despite a preponderance of experience and scientific advice to the contrary, Australia was still clearing native woody vegetation at something like three times the rate at which they replanted trees to solve the problems already emerging from earlier clearing.

Nevertheless, attitudes are changing. *In part, this is due to a growing recognition that trees and shrubs are an important functional component of many rangeland ecosystems, and in part it is due to the increasing value placed on landscapes for their environmental qualities.* While the latter may be an affordable luxury only for developed nations, the importance of the former underpins the sustainability of many rangelands for the benefit of their inhabitants worldwide.

While woody vegetation is perhaps a hindrance to the short-term maximization of high-input agricultural production systems, the importance of trees for fuelwood and other uses to low-input rangeland managers cannot be overstated. In this sense, the importance of trees has not changed, but rather the awareness that these resources must be used sustainably has increased as has the pressure on this resource.

Finally, there is a strong desire to improve the nutrition and economic well-being of rangeland communities in developing countries. The development of horticulture in areas traditionally too arid for tree crops is seen as a desirable diversification in land use and agricultural practice in many rangeland regions of Africa and Asia.

The Changing Views on Water in Rangelands

The general emphasis in rangeland management has, to date, been on *land* management, in support of primary production from that land. *The coincidence of increasing pressure for sustainable water resource development and the provision of water for environmental purposes has diminished the importance of rangeland primary production relative to the environmental services provided by rangelands, especially water.* Our improved understanding of the link between the way we manage vegetation and the quality and quantity of water flowing from the land adds impetus to considering how land management impacts on water resources.

Why Retain or Plant Trees in Rangelands?

Woody perennial vegetation is a natural feature of most rangelands around the world. However, in many regions it has been extensively cleared for grazing or agriculture, or simply through unsustainable wood harvesting (e.g. Australia, Chile, the Sahel). Such clearing is often associated with serious land degradation (Walker *et al.*, 1993; Hatton, 1999), and the need to restore or protect trees in the landscape is often driven by conservation needs (Campbell, 1994; Al-Qudah and Sabet, 1999; Christiansen and Wolff, 1999; Clary, 1999; Hatton and Nulsen, 1999; Ismail, 1999).

Alternatively, the impetus for tree planting is often economically driven. The need to supply fibre, food and fuelwood is great, and tree planting is being promoted and facilitated around the world in response to this need (Al-Laham, 1999; Berliner, 1999; Regner and El-Mowelhi, 1999; Sauerhaft *et al.*, 1999; Vertessy and Bessard, 1999). Such planting is either in support of new products or to sustain the inherent productivity of rangelands.

How Do Trees Affect the Water Resources of Rangelands?

There is evidence from Australia (Specht, 1972; Hatton and Wu, 1995; Ellis *et al.*, 1999), Spain (Joffre and Rambal, 1993) and the USA (Eagleson and Segarra, 1985) that in water-limited environments, the equilibrium (undisturbed) Leaf Area Index (LAI) of woodlands, shrublands and heaths is a function primarily of climate. Further, at equilibrium LAI, the vegetation minimizes runoff and groundwater recharge (Eagleson, 1982). This concept, if we accept it, offers a powerful paradigm for understanding and interpreting the hydrological consequences to afforestation or reforestation of rangelands. For instance, we can expect the downward displacement of the LAI–climate curve for a particular system (i.e. tree clearing) to result in greater water yield and perhaps groundwater discharge from the catchment.

Zhang *et al.* (1999) quantified one empirical expression of this idea for a large set of forested and grassland sites around the world (Fig. 14.1). In this case, the annual runoff from forested and non-forested catchments bore strong and distinct relationships to annual rainfall. What is harder to predict in such general terms are the conservation impacts of clearing-induced changes to the hydrological

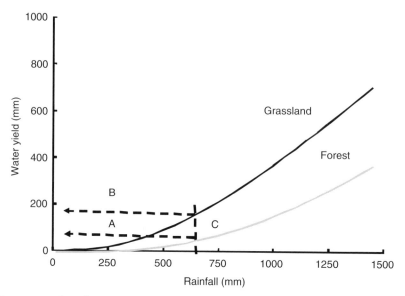

Fig. 14.1. The relationship between annual rainfall and catchment water yield, under grassland and forest (from Zhang *et al.*, 1999). In high rainfall country, where the concern is over streamflow reductions due to afforestation, this is a useful and robust relationship for estimating impacts. For instance, at 700 mm of rainfall (C), a grassland or previously cleared site may yield about 190 mm of runoff, while a forested catchment might yield only 60 mm; intermediate values of afforestation would result in a proportional decrease between these two yield values.

cycle. These can include decreased biodiversity, increased erosion, increased flooding, and salinization.

For instance, widespread clearing of native sclerophyll vegetation across southern Australia is resulting in a large dryland salinity problem (Hatton, 1999). Driven by the desire to reverse these hydrological impacts, tree planting is being promoted on a massive scale, with an acknowledgement that unless something like 10 million ha are reforested, much of the anticipated salinization of water resources will be realized. However, there is growing concern that if the siting of these new plantations is not appropriate, the decreases in stream flow will offset any benefits to reduced groundwater discharge and associated salt loads in rivers (Vertessy and Bessard, 1999). The Republic of South Africa has resorted to a water allocation policy requiring an allocation permit for tree plantations due to their impacts on catchment water yield (van der Zel, 1995).

A corollary to the ecohydrological equilibrium hypothesis as stated above is that one can make trade-offs between the density of a tree or stand canopy and the area of catchment supplying water to the trees (Fig. 14.2). Ellis *et al.* (1999) show how tree belts adjust their canopy

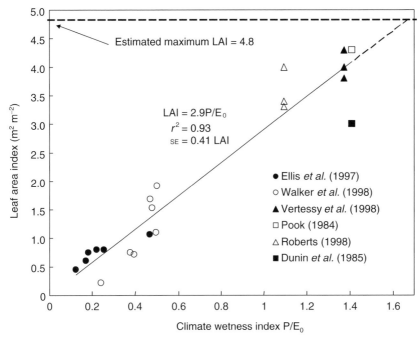

Fig. 14.2. LAI of natural eucalypt forest in southern Australia versus a climate wetness index; average annual rainfall P divided by average annual pan evaporation E_0. Displacement of LAI below this line (i.e. tree clearing) will likely result in increased groundwater recharge or runoff in rangeland environments (from Ellis *et al.*, 1999).

densities as a function of the equilibrium LAI of the site and the additional area providing water beyond the edge of the canopy. This is the principle behind run-on agroforestry as well: that even in the most arid zones otherwise hostile to tree survival or growth based on the expected relationship between LAI and climate, by harvesting rainfall over a larger area and focusing it on an area of limited canopy, trees can be made to grow.

Optimizing the degree of tree cover to manipulate rangeland water balance is not at all a modern idea. While tree cover has been dramatically reduced by people in most of the Mediterranean rangelands, the dehesas of the Iberian Peninsula are traditional (even ancient) rangeland systems that are managed to a tree density of 40–50 trees ha^{-1} for the production of grass and sweet acorns for people and livestock (Joffre and Rambal, 1993). More intriguingly, it appears that this density of trees is less than that associated with the equilibrium LAI, and the water resources released serve to reduce drought risk in the longer term.

Thus, the local water balance can be augmented through the clearing of the native woody vegetation, but often at the expense of biodiversity, soil loss and salinization; these impacts can in principle be reversed or limited by restoring trees to such landscapes. More local use of water (at the expense of downstream users) can be made by intercepting or redirecting runoff (natural or otherwise) to trees to maintain or enhance them in an environment otherwise too arid. Ultimately, there may be levels of tree cover that balance productivity and water resources in a sustainable and optimal way.

Who are the Stakeholders in Rangeland Water Resources?

It is important to realize that no water arising from rangelands is ever 'wasted'. Even if surface water or groundwater resources are not used or wanted by industry, these flow regimes inevitably support natural ecosystems that have inherent and in most cases tangible values (Hatton and Evans, 1998). Diversion of these water resources for tree growing must come at some cost downstream. The same may be said for the degradation of water resources through the imprudent clearing of trees.

The potential downstream stakeholders include water-driven industry, domestic users of surface water or groundwater, people who live and depend on floodplains, and a growing number of people concerned about conservation values like the preservation of biodiversity. In countries like Australia, the Republic of South Africa and the USA, these stakeholders comprise the vast preponderance of influence and ownership of rangeland water resources, as evidenced by water legislation and policy protecting their interests.

Therefore, issues like tree planting involve many considerations beyond the farm gate or communal lands, particularly with respect to the impacts on the catchment. The difficulty of resolving these potentially conflicting interests is hugely magnified when water resource impacts extend across scales (Weltz *et al.*, 1999) or political boundaries (e.g. Seely *et al.*, 1999). In the case of Australia (a federation of states with largely independent policies on catchment management), the Murray–Darling basin extends across six states and territories. This catchment currently produces about half of Australia's food, but is under intense pressure due to the over-allocation of water, the salinization of rivers and the need for rural development. Tree planting is advocated in a variety of forms and for a range of purposes that would produce significant changes to the water resources of the Murray–Darling basin (Vertessy and Bessard, 1999). The issue of social and economic equity in the shape of these changes demanded the development of an interstate instrument (the Murray–Darling basin

Fig. 14.3. Secondary salinization resulting from clearing of native woodlands in Western Australia. If the solution, or containment, of this problem lies in restoring tree cover to something like the original extent and density, who pays and who benefits?

Commission) with wide-ranging powers to manage such issues (Fig. 14.3).

Tree Planting in Rangelands: Who Wins?

The key to sorting out winners and losers in the issue of tree planting is in the identification of on-farm (or alternatively, tribal or communal) benefits and costs, and the benefits and costs in terms of environmental, economic and social services delivered to the wider community from rangeland catchments.

Where tree planting is driven largely by the conservation needs of the wider community, it may be that there are commercial forestry or agroforestry opportunities that can compete with traditional land use on an economic basis and fit in well with local values and aspirations. This is the ideal ('win–win'), and adoption is usually only a matter of extension of information and perhaps some initial financial support from government or developers. In Australia, the need to afforest vast areas to control salinization is only likely to be met if done on a commercial basis, and the development of the oil mallee (a eucalyptus tree yielding multiple products including oil, activated charcoal and fuel biomass) and maritime pine industries for regions with less than 600 mm of annual rainfall may achieve commercial parity with conventional farming.

In most cases, however, conservation goals will not be met through commercially viable forestry or agroforestry (e.g. McDaniel and Taylor, 1999). In these cases, the motivation for tree planting by local rangeland managers is limited to their sense of goodwill and degree of conservation ethic. Given the declining terms of trade for many primary producers (Hatton, 1999) or their historic levels of poverty, the collective ability to finance conservation tree planting from among local rangeland managers is generally quite limited. Additionally, the restoration of water quality through conservation-driven tree planting may come at the cost of lost opportunity for landholders desiring higher value uses.

Where the imperative for tree planting is driven by local economic need (fuelwood, industry development, increasing sustainability), and local economics cannot wholly drive tree planting initiatives, then the wider community must participate in some way. *The cost-sharing arrangements for this balancing of private good and public need require input from all stakeholders* (Barber 1999; Pegler *et al.*, 1999). Such decisions are complex and must take into account not only the downstream impacts on water resources but also the regional or national desires for social and economic equity (e.g. Al-Laham, 1999; Al-Qudah and Sabet, 1999; Brake *et al.*, 1999; Ismail, 1999; Regner and El-Mowelhi, 1999). Syme *et al.* (1999) provide a discussion of instruments to facilitate the broader discussion of stakeholder equity and community goal setting for integrated catchment management issues like tree planting.

Conclusions

The management of tree cover (clearing native woody vegetation, commercial forestry, agroforestry) in rangeland catchments has a profound effect on the yield and quality of water. This dependency of water on tree management creates the potential for conflict among the wider stakeholders dependent on the flows of water of a particular quality. The breadth of these stakeholders may not be immediately obvious to those making decisions about tree management, and includes not only downstream users of water but also those members of the community with interests in conservation, cultural continuity, social justice and economic equity.

Resolving such conflicts depends on first identifying and then consulting stakeholders in regard to the benefits and impacts of the proposed vegetation management. This may include the provision of technical advice and education to underpin the articulation and quantification of concerns. Once the equity in the issue is identified, decisions regarding cost sharing between the wider community and local rangeland managers can be agreed.

Trees have the potential to greatly enhance the value of rangeland to the local and downstream communities. There are compelling ecohydrological reasons for a given degree of tree cover with respect to the stability and sustainability of landscapes. These principles are useful in designing or redesigning forestry and agroforestry systems for the long-term benefit of people in rangelands.

References

Al-Laham, A. (1999) Trust-building and partnership between authorities and local communities in Palestinian degraded rangelands. In: Eldridge, D. and Freudenberger, D. (eds) *People of the Rangelands. Building the Future. Proceedings of the VI International Rangeland Congress.* VI International Rangeland Congress Inc., Townsville, Australia, p. 692.

Al-Qudah, B. and Sabet, J. (1999) Rangeland management, water harvesting and users' involvement with rangeland and conservation reserves in Jordan. In: Eldridge, D. and Freudenberger, D. (eds) *People of the Rangelands. Building the Future. Proceedings of the VI International Rangeland Congress.* VI International Rangeland Congress Inc., Townsville, Australia, pp. 690–691.

Barber, D.R. (1999) Integrating resource information for regional planning, vegetation and water reform in the Far West region, New South Wales, Australia. In: Eldridge, D. and Freudenberger, D. (eds) *People of the Rangelands. Building the Future. Proceedings of the VI International Rangeland Congress.* VI International Rangeland Congress Inc., Townsville, Australia, pp. 704–705.

Berliner, P. (1999) Run-off harvesting for firewood and fodder production in Israel. In: Eldridge, D. and Freudenberger, D. (eds) *People of the Rangelands. Building the Future. Proceedings of the VI International Rangeland Congress.* VI International Rangeland Congress Inc., Townsville, Australia, p. 693.

Brake, L., Sorensen, B., Ireland, C. and Power, N. (1999) The community and water management in the arid areas of South Australia. In: Eldridge, D. and Freudenberger, D. (eds) *People of the Rangelands. Building the Future. Proceedings of the VI International Rangeland Congress.* VI International Rangeland Congress Inc., Townsville, Australia, pp. 705–706.

Campbell, A. (1994) *Landcare.* Allen and Unwin, St Leonards, New South Wales, Australia, 344 pp.

Christiansen, S. and Wolff, H.P. (1999) International Middle East collaboration in rangeland management – the initiative for collaboration to control natural resource degradation. In: Eldridge, D. and Freudenberger, D. (eds) *People of the Rangelands. Building the Future. Proceedings of the VI International Rangeland Congress.* VI International Rangeland Congress Inc., Townsville, Australia, pp. 687–688.

Clary, W.P. (1999) Riparian-stream habitat responses to late spring cattle grazing. In: Eldridge, D. and Freudenberger, D. (eds) *People of the Rangelands. Building the*

Future. Proceedings of the VI International Rangeland Congress. VI International Rangeland Congress Inc., Townsville, Australia, pp. 695–697.

Eagleson, P.S. (1982) Ecological optimality in water-limited natural soil–vegetation systems. 1. Theory and hypothesis. *Water Resources Research* 18, 325–340.

Eagleson, P.S. and Segarra, R.I. (1985) Water-limited equilibrium of savanna vegetation systems. *Water Resources Research* 21, 1483–1493.

Ellis, T., Hatton, T.J. and Nuberg, I. (1999) A simple method for estimating recharge from low rainfall agroforestry systems. In: *ENVIROWATER99 2nd Inter-Regional Conference on Water–Environment Emerging Technologies for Sustainable Land Use and Water Management.* Lausanne, Switzerland.

Hatton, T.J. (1999) Multiple issues in water and land management in Australian catchments. In: Eldridge, D. and Freudenberger, D. (eds) *People of the Range-lands. Building the Future. Proceedings of the VI International Rangeland Congress.* VI International Rangeland Congress Inc., Townsville, Australia, pp. 675–678.

Hatton, T.J. and Evans, R. (1998) Dependence of Australian ecosystems on groundwater. *Water* 25, 40–43.

Hatton, T.J. and Nulsen, R.A. (1999) Towards achieving functional ecosystem mimicry with respect to water cycling in southern Australian agriculture. *Agro-forestry Ecosystems* 45, 203–214.

Hatton, T.J. and Wu, H. (1995) Scaling theory to extrapolate individual tree water use to stand water use. *Hydrological Processes* 9, 527–540.

Ismail, M. (1999) Runoff harvesting for in-situ conservation and use of range, forest and horticultural germplasm in Tunisia. In: Eldridge, D. and Freudenberger, D. (eds) *People of the Rangelands. Building the Future. Proceedings of the VI International Rangeland Congress.* VI International Rangeland Congress Inc., Townsville, Australia, p. 691.

Joffre, R. and Rambal, S. (1993) How tree cover influences the water balance of Mediterranean rangelands. *Ecology* 74, 570–582.

McDaniel, K.C. and Taylor, J.P. (1999) Steps for restoring bosque vegetation along the middle Rio Grande of New Mexico. In: Eldridge, D. and Freudenberger, D. (eds) *People of the Rangelands. Building the Future. Proceedings of the VI International Rangeland Congress.* VI International Rangeland Congress Inc., Townsville, Australia, pp. 713–714.

Pegler, L., Rose, A. and Bentley, D. (1999) Changed water distribution in Australia's rangelands: economic, ecological and social perceptions of land managers. In: Eldridge, D. and Freudenberger, D. (eds) *People of the Rangelands. Building the Future. Proceedings of the VI International Rangeland Congress.* VI International Rangeland Congress Inc., Townsville, Australia, pp. 701–702.

Regner, H.J. and El-Mowelhi, N. (1999) Support to Bedouin communities for soil erosion control and water conservation for a comprehensive rural development in the north-west coast of Egypt. In: Eldridge, D. and Freudenberger, D. (eds) *People of the Rangelands. Building the Future. Proceedings of the VI International Rangeland Congress.* VI International Rangeland Congress Inc., Townsville, Australia, pp. 688–690.

Sauerhaft, B.C., Thurow, T. and Berliner, P. (1999) Spacing and green manure effects on biomass yield in an aridland water catchment agroforestry system. In: Eldridge, D. and Freudenberger, D. (eds) *People of the Rangelands. Building the*

Future. Proceedings of the VI International Rangeland Congress. VI International Rangeland Congress Inc., Townsville, Australia, pp. 697–698.

Seely, M.K., Jacobson, K.M., Jacobson, P.J., Leggett, K. and Nghitila, T. (1999) Understanding integrated natural resource management at a catchment scale: the case of Namibia's ephemeral rivers. In: Eldridge, D. and Freudenberger, D. (eds) *People of the Rangelands. Building the Future. Proceedings of the VI International Rangeland Congress.* VI International Rangeland Congress Inc., Townsville, Australia, pp. 714–716.

Specht, R.L. (1972) Water use by perennial evergreen plant communities in Australia and Papua New Guinea. *Australian Journal of Botany* 20, 273–299.

Syme, G.J., Butterworth, J.E. and McCreddin, J.A. (1999) Living with integrated catchment management. In: Eldridge, D. and Freudenberger, D. (eds) *People of the Rangelands. Building the Future. Proceedings of the VI International Rangeland Congress.* VI International Rangeland Congress Inc., Townsville, Australia, pp. 683–687.

van der Zel, D.W. (1995) Accomplishments and dynamics of the South African afforestation permit system. *South African Forestry Journal* 176, 55–59.

Vertessy, R.A. and Bessard, Y. (1999) Conversion of grasslands to plantations: anticipating the negative hydrologic effects. In: Eldridge, D. and Freudenberger, D. (eds) *People of the Rangelands. Building the Future. Proceedings of the VI International Rangeland Congress.* VI International Rangeland Congress Inc., Townsville, Australia, pp. 679–683.

Walker, J., Bullen, F. and Williams, B.G. (1993) Ecohydrological changes in the Murray–Darling Basin. I. The number of trees cleared over two centuries. *Journal of Applied Ecology* 30, 265–273.

Weltz, M.A., Kidwell, M.R., Fox, H.D. and Yakowitz, D. (1999) Assessing scale issues on semi-arid watersheds with a multi-attribute decision. In: Eldridge, D. and Freudenberger, D. (eds) *People of the Rangelands. Building the Future. Proceedings of the VI International Rangeland Congress.* VI International Rangeland Congress Inc., Townsville, Australia, pp. 694–695.

Zhang, L., Dawes, W.R. and Walker, G.R. (1999) *Predicting the Effect of Vegetation Changes on Catchment Average Water Balance.* Cooperative Centre for Catchment Hydrology Technical Report 12/99, Monash University, Clayton, Victoria, Australia, 44 pp.

Land and Water Management: Lessons from a Project on Desertification in the Middle East

15

Scott Christiansen

Introduction

A diplomatic and technical support structure exists to help resolve the conflict in the Middle East, which was established at the Madrid Peace Conference in 1991; a process infrequently described, but well summarized for its first 5 years in a book by Joel Peters (1996). The multilateral talks were devised to provide a multilateral forum for discussion of non-political trans-boundary issues of concern to all parties that would independently build confidence through dialogue and stimulate the discussion of country-to-country issues among the front-line states.

The basic structure of the process, which continues to this date, includes five different Working Groups (arms control, regional economic development, refugees, water and the environment). One project in the Working Group on the Environment (WGE) was designed to control natural resource degradation in Egypt, Israel, Jordan, the Palestinian Authority and Tunisia. The project had a large focus on rangeland, which was but one of many projects in the WGE; nevertheless, it can be safely stated that this project reflects some of the challenges faced by the entire Middle East peace process.

It is clear that a regrouping is now called for that would revisit the successes and failures of the Peace Talks with a view to more widely publicize the next iteration of the process to achieve a wider public awareness and acceptance of the effort.

©CAB International 2002. Global Rangelands: Progress and Prospects (eds A.C. Grice and K.C. Hodgkinson)

Project Background

The World Bank was instrumental in facilitating the process of preparation and implementation of the project, which was entitled the 'Initiative for Collaboration to Control Natural Resource Degradation (Desertification) of Arid Lands in the Middle East' recently renamed the 'Regional Initiative for Dryland Management' for Phase II of the project, and scheduled to continue through to 2003. The World Bank, Switzerland, Luxembourg, Japan, the USA, the European Commission, the Republic of Korea and Canada provided funding that is complemented by in-kind and financial inputs by the participating countries. The International Centre for Agricultural Research in the Dry Areas (ICARDA) is responsible for programme execution through a Facilitation Unit hosted in the ICARDA-Cairo offices.

The project is unique as a pilot project because, if successful, the idea could be used elsewhere around the globe where political conflict and environmental degradation are unhappily brought together and where cooperation might yield a solution to both environmental and political problems. By 1999, a relatively cold Middle Eastern political atmosphere had thawed; however, during the writing of this book the Middle East conflict again entered another age of political stalemate.

Figure 15.1 shows the original design for the implementation of Phase I (1996–2000). Countries leading Regional Support Programmes (RSPs) are indicated with shading. Each country carried out activities for each theme within its territory – called National Support Activities (NSAs). In addition, the Palestinian Authority was assisted in capacity building through support to the establishment of the Palestinian Environmental Authority, which was subsequently merged with the Ministry of Environmental Affairs. For example, Jordan was to carry out NSAs within its territory and also provide regional support to all the other countries in the RSP for the Rangeland Management theme. NSAs were planned as research and development projects while the RSPs are used for coordination, training and communication.

The original project structure required a donation of funds from a variety of donors – some of them understandably reluctant to invest in a political process that was going up and down like a roller-coaster. As a result, Phase I funding was adequate but unevenly distributed. Regional work was generously funded by the Swiss and included Israel, but a freeze on multilateral activities called by the Arab League from March of 1997 until 2000 limited the extent of regional work.

Meanwhile, no donors or participating countries came forward with funding to support national activities in Egypt, Israel or Tunisia, which made it necessary to repeatedly reallocate small amounts of World Bank funds designed to operate the Facilitation Unit to Egypt and Tunisia with the hope that a donor would soon contribute

Regional Suppport Programme (RSP)	National Support Activity (NSA)				
Theme	Egypt	Israel	Jordan	Tunisia	Palestinian Authority
Germplasm for Arid Lands	NSA & RSP Leader	NSA	NSA	NSA	NSA
Germplasm for Arid Lands	NSA	NSA & RSP Leader	NSA	NSA	NSA
Economic Forestry and Orchards	NSA	NSA	NSA & RSP Leader	NSA	NSA
Rangeland Management	NSA	NSA	NSA	NSA & RSP Leader	NSA
Capacity Building	– –	– –	– –	– –	Focal Point

Fig. 15.1.   Original design for the implementation of Phase I (1996–2000).

resources. Israel's high per capita income made it ineligible for the overseas development funding of most donors, including the World Bank, and the Ministry of Foreign Affairs in Israel did not contribute funds for their own NSAs in Phase I. As a result of the above circumstances, the project was often forced to justify the use of regional funds for national activities with the chicken-and-egg logic that if there were no funds to develop national activities then there would be nothing to show during the regional tours and demonstrations.

Most of the national funds were concentrated within Jordan and the Palestinian Authority. The Palestinians had funds for all four NSAs but limited institutional capacity, and competition for leadership of the project caused slow disbursement of funds. The Jordanians had three of four NSAs funded within the project; however, donors outside the project were also funding dryland activities that in turn limited the need to disburse funds allocated by the Initiative.

On top of the funding imbalances and differences in rates of disbursement, efforts to cooperate technically between Israel and the Arab partners in the project were continually affected by the changing political circumstances in the Middle East. Eventually, project implementation strategies were created to accommodate political contingencies through a flexible expansion and contraction of activities that could be politically accepted at any given time.

The biggest contribution of Phase I, between 1996 and 2000, was to call for an External Review that restructured the project so that each country had national project activities, which were formally funded as an integral part of the project, where work could contract but continue during difficult political periods. Regional efforts thus become incremental to national efforts, adding regional value to existing work.

By the end of Phase I in 2000, idealism gave way to pragmatism and the hypothesis that regional exchanges can make headway through technical cooperation independently of political progress was disproved.

Importantly, the project is now simplified, focused on fewer physical locations and shifted more towards development than research. Cooperation during Phase II focuses on: (i) natural resource conservation and watershed management using water harvesting techniques and suitable plant resources; (ii) treated wastewater and bio-solids management, reuse and guidelines; and (iii) identification of policy options for reversing natural resource degradation.

Lessons Learned

1. Development objectives are retarded when politics prevent project activities from being implemented. Division and vacillating support at the donor level for management and organization of the project bear part of the burden for failures. In the financial context, slow disbursement or reporting slows the entire project cycle. In multi-institutional settings the communication among institutions can become contentious. In resource-limited multidisciplinary settings, competition may limit cooperation. Within institutions, key coordinators can often keep essential information from being shared, limiting the opportunity for debate and discussion. The scientists sometimes lack initiative or responsibility, particularly when decisions are political, i.e. reserved for top managers directly linked to political supervisors. Finally, when politics are not at issue, an inadequate incentive system and incomplete delegation of authority impedes progress.

2. In situations where organization and management are properly functioning, a problem of all research or development projects is to exclude beneficiaries in the design and implementation of projects. Exclusion of farmer stakeholders perpetuates a system whereby governments, regional and international organizations can overly influence the targeting of donor assistance. Donors should insist on creating balanced and complementary views of development through National Management Committees for each country, which must include representation of the target population affected by desertification. There is no rule for composition of a National Management Committee; however, donors should have enough leverage over the selection process to ensure a representative cross-section of stakeholders, one in which healthy competitive forces generate open debate and compromises, particularly with reference to the share of resources spent on solving problems for the users that are affected by desertification.

A generalized composition for the committee is recommended as follows:

- Ministry of Environment (conservation)
- Ministry of Agriculture (production)
- Ministry of Water and/or Irrigation (hydrology, treated wastewater)
- A relevant university (advanced training, specialized studies)
- A non-government organization (independent funding possibilities)
- Representative from the community or Civic Society (local perspective).

3. Land ownership and user rights are topics of immense proportions that have obvious, but not often addressed, relevance to the vast areas of rangeland throughout North Africa, the Middle East and Central Asia. Policy decisions to grant rights of use should start with simple pilot demonstrations so that governments can take a step-by-step approach in finding the right model for reform. Differences in the land use tenure in the countries of the region make some technologies less transferable than others across countries.

4. In very dry settings unfit for cultivated crops, fodder shrubs or fruit trees, combined with water harvesting, can be managed to produce functional and profitable agricultural systems. Some scientific issues for the production of fodder shrubs include:

- growing shrubs in wide-spaced rows within barley (Le Houérou *et al.*, 1991);
- formulating balanced rations with shrubs and other available feeds and phasing shrubs gradually into animal diets (Le Houérou, 1991b); and
- potential to put the shrubs into feed blocks, which provides an incentive for farmers who do not own livestock to grow shrubs for production of a commodity that can be stored, transported and marketed.

5. Water of marginal quality is the only water resource that is increasing in availability. Standards are needed to regulate the use of treated wastewater and bio-solids, and methods are needed to improve the treatment so as to decrease health risks when reusing these valuable agricultural resources. It is critical in many areas where fresh water is scarce to save it for domestic use. Reuse strategies and monitoring systems are needed to use lower quality water in agricultural production and development of dryland zones.

6. During the multilateral freeze imposed by Arab states on cooperation with the Israelis between 1997 and 2000 there were very few, if any, interactions among Working Groups within individual countries. A more flexible and innovative project structure would have offered an opportunity for the individual projects within each of the developing Arab states to continue their work and maintain their momentum during times of political impasse. This is linked to the idea of having sustainable national development projects as the basis for Arab–Israeli cooperation in the future.

7. The structure of the Multilateral Middle East Peace Talks is comprehensive and philosophically interesting. Practically speaking, however, it is too complicated. To even the most fastidious students of the process as a whole, it has lost its coherence to the point where communication among its layers of organization has broken down – within and across projects and Working Groups – upwards to the Steering Committee. It is for this reason that it is believed that a thorough simplification and streamlining of the organizational structure and a serious reconsolidation of the process is now in order.

A New Global Dryland Setting

A new role for developed countries will be to help preserve peace, address issues of globalization and to improve the quality of life of the many poor and disadvantaged throughout the world (Lancaster, 2000). As the world continues to make unprecedented demands upon natural resources, one major responsibility will be to develop partnerships among countries to improve management of land and water resources. This is very much what is being done in the Regional Initiative for Dryland Management. If this work is to be expanded to other areas of the world it will be important to consider trends in funding for agricultural research and development.

History: post-Second World War agricultural development evolution

International development assistance is a post-Second World War phenomenon that was given momentum by the Point Four Program, which was the US policy of technical assistance and economic aid to underdeveloped countries, so named because it was the fourth point of President Harry S. Truman's 1949 inaugural address. Emphasis was placed on technical assistance, largely in the fields of agriculture, public health and education, furnished through specialized United Nations agencies, as well as through US contributions on a bilateral basis, often channelled through land-grant universities. It is important to remember that this was a time when investments were often made to affect the strategic balance of power and competing ideologies between the free-market 'Western' world and the Soviet-dominated command economies.

The 1950s, 1960s and 1970s were a time of large multi-sector investment projects, heavy on support to irrigation, roads, infrastructure – an era of 'bricks and mortar'. Livestock and rangeland projects addressed issues such as drilling of wells for livestock watering points, veterinary care and subsidized feeds, re-vegetation of degraded rangeland and establishment of feed reserves with fodder shrubs, creation of pastoral fattening cooperatives and settling of nomads (El-Shorbagy, 1998).

Rangeland has been less important in feeding the world than crops. In the 1970s, the Green Revolution produced improved varieties of rice at the International Rice Research Institute (IRRI) and wheat at the International Maize and Wheat Improvement Centre (CIMMYT). These technology packages also called for fertilizer and management packages to sustain high yields. In part due to the successes of IRRI and CIMMYT, a succession of International Agricultural Research Centres (IARCs) continued to be born into the 1990s under the auspices of the Consultative Group for International Agricultural Research (CGIAR). Eighteen IARCs were created before a consolidation brought the number down to 16 Centres. The World Bank and USAID were the largest contributors to the CGIAR so it is not surprising that many of the Green Revolution crop commodity concepts were adopted and replicated wholeheartedly around the developing world.

Research and development projects brought new financial resources, equipment and technology to all corners of the world. This trend can be correlated with the unintended but inexorable loss of rangelands by conversion to cropland. Taking the Mediterranean region as an example, over the past 50 years, about 50% of arid rangelands between the 100–400 mm bands of mean annual rainfall have been cleared, mostly for cultivation of barley (Le Houérou, 1991a). These were the most productive rangelands with relatively deep soils, located in topographic depressions, benefiting from some rainfall run-on. Over the past decades, bad management has converted the best rangelands to some of the worse cropland, often causing desertification and abandonment of these lands.

The 1980s were a period of donor fatigue, recession and stock market slumps that simultaneously trimmed spending on development. The World Bank spent US$2 billion on range and livestock activities from 1974 to 1979, which fell to less than US$50 million in the period 1980–1985 (Niamir-Fuller, 1999). In the 1990s, the research agenda was broadened, shifting to multidisciplinary or systems approaches including farming systems, socio-economics, participatory processes, gender studies and the inclusion of natural resources in the research and development agenda. Rangelands have subsequently been more or less embedded within natural resources. Foran and Howden (1999) contend that rangelands will continue their decline in world and national affairs. Although the areas are vast, their inhabitants are few. Most project personnel are not well adapted to these settings and their offices are back in the cities where the majority prefers to live comfortably with access to civilization and services. Programmes are needed that will bring self-sufficiency to those who want to live on the range, for personal, political or cultural reasons.

In the larger geopolitical arena, attention turned to the collapse of the communist system and many funds were diverted to encourage new paths towards democratic and market-oriented reform. However, in 1992, the United Nations Conference on Environment and Development (UNCED) developed the 'Rio treaties' – all three of which are of direct concern to the future of rangeland development: the Framework Convention on Climate Change (FCCC), the Convention on Biological Diversity (CBD) and the Convention to Combat Desertification (CCD). Through the Global Environment Facility, new funding has been made available to restart development in the rangelands based on an environmental approach as opposed to one that focuses on production (Lusigi and Acquay, 1999).

Technical and Organizational Trends

Ultimately, natural resource management is the result of interaction among three quite distinct systems – resources, resource users, and the larger geopolitical system in which they operate – each of which needs to be understood if intervention is to be effective (Pratt *et al.*, 1997). As we enter the new millennium, the need to predict and gauge the direction of research and development will not change. The future requires organization and technology so that the best new ideas can be put into practice as soon as possible. What is likely to remain in the forefront of rangeland science during the next decade will surely include combinations of topics discussed in the remainder of this chapter.

Agriculture and environment

During the past decade, support for agriculture declined while support for the environment went up. Before long, it became clear that both sectors were inextricably intertwined. While resources are targeted at environmental issues, the world recognizes that agriculture is both a cause of environmental degradation as well as a solution for sustainable environmental management.

Carbon sequestration

Evidence that the atmospheric build-up of greenhouse gases from anthropogenic sources is causing global climate change has created new opportunities for the energy sector to produce cleaner fuels and reduce greenhouse gas emissions. Rejuvenation of funding opportunities is anticipated in the agriculture and environment sectors through sequestration of carbon dioxide in rural landscapes. One of the big promises for rural development may be the establishment of grassland conservation programmes to rehabilitate the degraded lands, sequestering and storing significant quantities of organic matter, and thereby offering disadvantaged rural farmers an opportunity to benefit from strategies designed to mitigate climate change through trading of carbon credits with industrialized countries.

Safeguarding and using our waters

Reclaiming municipal wastewater and bio-solids for agricultural re-use will be high on the global agenda for research and development alike. Wastewater re-use improves the environment because it reduces the amount of waste discharges into watercourses, and it conserves water resources by lowering the demand for freshwater abstraction. In the process, re-use has the potential to reduce the costs of both wastewater disposal and the provision of irrigation water, mainly around cities and towns with sewers at present, but more and more in rural settings, particularly dry ones, where every drop of water is a precious commodity.

Remote sensing and decision-support tools

There is a critical need for improved technology to develop integrated ranch and catchment management systems that are environmentally sustainable and economically viable. Trans-boundary catchments and rangelands are a shared resource calling for standard approaches and common databases in decision-making. Countries must share common goals to understand the basic processes that affect rangelands and catchments, and develop and implement management strategies that will sustain our resources for future generations. For example, it is already possible to use spatially explicit hydro-ecological models calibrated with satellite images to produce daily estimates of regional plant growth and rangeland health that are three times more accurate than conventional methods without satellite imagery. It will soon be possible to measure soil moisture at a large scale using airborne and space microwave platforms. Soil moisture is a critical variable for climate and agriculture, determining the partitioning of water and energy between the earth's surface and atmosphere, and impacting on net primary productivity and weather. Up until now, measuring soil moisture over continental scales was hindered by appropriate instrumentation.

Regional linkages among donor programmes

Many governments have programmes of bilateral assistance to developing countries. Often these efforts are similar and focused on the same target ministry in the developing country. There are some instances where development assistance is addressed by adjacent countries in the same eco-geographical setting where clear advantages of cooperation could be achieved through a reconciliation of the programmes. Where research and development needs can be expensive, particularly in natural resource management, regional cooperation is clearly needed.

Back in 1992 the world gathered in Rio de Janeiro to map out an environmental agenda for the 21st century that took into consideration land degradation, loss of biodiversity and global climate change. This 'Agenda 21' calls for good demonstrations to show how dryland countries can achieve environmental and economic stability. If donor nations could lend their support to the coordinated implementation of regional action programmes, then successful activities can be replicated and used to meet national obligations for other countries that have ratified and have become parties to the FCCC, CBD and CCD.

As the UNCCD now accelerates the establishment of its Sub-Regional Action Plans, the projects that are created should strive for good governance and a focus on problem solving. The projects should monitor progress and expected outputs, and include an information management system that is accessible by all parties through the Internet. An honest and transparent system of monitoring by peers can help to solve problems amicably at a local level to prepare project teams for the inevitable project evaluation; however, participants and donors alike should accept that independent outside reviews are an excellent way to introduce fresh views at critical points in the life cycle of all projects.

The role of the CGIAR's future harvest centres

The Consultative Group for International Agricultural Research (CGIAR) is in jeopardy. Every year they gather at Centre's Week, hosted at the World Bank in Washington, and donations are collected to run the system for another year. In the 2000 meetings the system renamed itself as Future Harvest, striving for a new corporate identity while maintaining coordination and traditional sources of financial support.

The Future Harvest Centres are worthy of support as they have a good name with the National Agricultural Research Systems (NARS) that they serve and there is a sound financial reporting capability that donors can rely upon for their records. It is calculated that the CGIAR system represents only 4% of all financial support to agricultural research but it is a very heavily leveraged support that is widely appreciated. With a consistent level of support the Future Harvest Centres could start to rely more strategically upon the strong NARS in each region. Brazil is a good example of a strong NARS in South America, which can be relied upon to transfer the best findings to other countries on the continent.

NARS, development banks and the CGIAR are critical elements in the system of global agriculture/environment research and development coordination. At the moment the CGIAR is operating on a very small budget but has the confidence of the NARS and provides financial accountability to donors. The development banks have the resources and development agenda but lack the latest technology and scientists for implementation of projects in rural development.

Lack of a larger vision

Debates over globalization have handicapped bold initiatives that could help to erase rural poverty in the 21st century. A trust fund could be generated to financially stabilize the CGIAR. Until this happens it seems that the Future Harvest Centres may not have a very bright future. If a sustainable funding system was created it would be relatively easy to ensure that the Centres rejuvenate themselves through obligatory term limits on appointments of 5–7 years. This is long enough to create solid outputs and would open a continuous, sustainable recycling of scientists from NARS and advanced research organizations.

Conclusions

As the UNCCD now accelerates the establishment of its Sub-Regional Action Plans around the world, projects that are created should strive for good governance and a focus on problem solving. As with the piloting efforts of the 'Regional Initiative for Dryland Management' in the Multilateral Middle East Peace Process, the UNCCD could generate a global network of Regional Facilitation Units for the dry areas of the world, backstopped by research from the Future Harvest Centres and advanced research organizations. When successes are achieved in this network they could be scaled up with assistance from governments and development banks. The projects should monitor progress and expected outputs and include an information management system that is accessible by all parties. An honest and transparent system of technical monitoring and financial auditing by competent external professionals could elevate work standards to those used within normal business practices.

As shown in the US, agricultural research has had an extraordinary return on research investments. Research and development are insepa-rable and it is time that the world made concrete plans to eradicate rural poverty. This lofty goal will require organization and sharing of knowledge through a promotion of comparative advantages in trans-ferring knowledge. The effort must combine research and development on natural resource management and engage countries to participate in Agenda 21 to achieve environmental and economic stability in the coming generation.

References

El-Shorbagy, M.A. (1998) Impact of development programs on deterioration of rangeland resources in some African and Middle Eastern countries. In: Squires, V.R. and Sidahmed, A.E. (eds) *Drylands: Sustainable Use of Rangelands into the Twenty-First Century*. IFAD, Rome, Italy, pp. 45–70.

Foran, F. and Howden, M. (1999) Nine global drivers of rangeland change. In: Eldridge, D. and Freudenberger, D. (eds) *People of the Rangelands. Building the Future. Proceedings of the VI International Rangeland Congress*. VI International Rangeland Congress Inc., Townsville, Australia, pp. 7–13.

Le Houérou, H.N. (1991a) Rangeland management in northern Africa and the Near East: evolution, trends and development outlook. In: Gaston, A., Kernick, M. and Le Houérou, H.N. (eds) *Proceedings of the IV International Rangeland Congress*, UCIST/CIRAD, Montpellier, Vol. 1, pp. 543–551.

Le Houérou, H.N. (1991b) Feeding shrubs to sheep in the Mediterranean arid zone: intake performance and feed value. In: Gaston, A., Kernick, M. and Le Houérou, H.N. (eds) *Proceedings of the IV International Rangeland Congress*, UCIST/CIRAD, Montpellier, Vol. 2, pp. 623–628.

Le Houérou, H.N., Correal, E. and Lailhacar, S. (1991) New man-made agro-sylvopastoral production systems for the isoclimatic Mediterranean zone. In: Gaston, A., Kernick, M. and Le Houérou, H.N. (eds) *Proceedings of the IV International Rangeland Congress*, UCIST/CIRAD, Montpellier, Vol. 1, pp. 383–388.

Lancaster, C. (2000) Redesigning foreign aid. *Foreign Affairs* 79(5), 74–88.

Lusigi, W.J. and Acquay, H. (1999) The GEF experience in supporting programs related to land degradation in developing countries. In: Eldridge, D. and Freudenberger, D. (eds) *People of the Rangelands. Building the Future. Proceedings of the VI International Rangeland Congress*. VI International Rangeland Congress Inc., Townsville, Australia, pp. 156–159.

Niamir-Fuller, M. (1999) International aid for rangeland development: trends and challenges. In: Eldridge, D. and Freudenberger, D. (eds) *People of the Rangelands. Building the Future. Proceedings of the VI International Rangeland Congress*. VI International Rangeland Congress Inc., Townsville, Australia, pp. 147–153.

Peters, J. (1996) *Pathways to Peace: The Multilateral Arab–Israeli Peace Talks*. Royal Institute of International Affairs, London. Distributed by the Brookings Institute, Washington, DC, 110 pp.

Pratt, D.J., Le Gall, F. and de Haan, C. (1997) *Investing in Pastoralism: Sustainable Natural Resource Use in Arid Africa and the Middle East*. World Bank Technical Paper No. 365. Washington, DC, 159 pp.

International Perspectives on the Rangelands

16

Wolfgang Bayer and Peter Sloane

Introduction

International perspectives on rangelands have multiple dimensions. These may be physical and can include issues which affect large parts of the rangelands, such as desiccation, frequent droughts, a decreasing number of people depending exclusively on rangelands, or crop encroachment. They may have spatial dimensions with regional or local focus. Many functionally linked rangeland areas are not defined by national boundaries, so their use is not solely an issue for one country or government. External assistance for rangeland development and management is another international dimension and is the focus of this chapter. Particular emphasis is given to the lessons learned over the past 40 years and to the extent to which these lessons brought about any changes in the approaches to assistance taken by donors, technical and implementing agencies (whether multi- or bilateral agencies or international non-governmental organizations), and provided signals as to future pathways for assistance to achieve sustainable development and management of the world's rangeland resources.

Evolution of Rangeland Development Concepts

Development projects involving interventions in the rangelands commenced about 40 years ago. The financial commitment has fluctuated over this period, with World Bank spending for interventions in the rangelands having fallen to a current level of US$2 million year^{-1} from a high point of US$20 million year^{-1} in the late 1970s and early 1980s, though it is now showing some signs of increasing again (de Haan, 1999). Trends in spending on rangeland interventions by other agencies have been similar. *Rangeland development is now often not a separate project, but part of a wider programme of natural resource management.*

Concepts in range development have also changed. The 1960s and early 1970s were the period of production and technology transfer (Turk, 1999). The main objective was to increase production for the urban and international market, and the approach was one of capital-intensive and labour-extensive ranching, with range restoration and improvement schemes, water-point development and fodder production (Niamir-Fuller, 1999).

The ranching approach failed for ecological, economic and institutional reasons. Local authorities administering the schemes could rarely enforce rules, nor could they prevent internal corruption. Where successful, these schemes undermined traditional rules and facilitated the intrusion of non-pastoralists. Projects such as state-sponsored and donor-assisted water-point development were intended to increase livestock production by opening up new pastures. Instead, however, *they often disturbed traditional patterns of rangeland use, transforming them into open-access situations, which accelerated degradation of the rangeland* (Niamir-Fuller, 1999; Turk, 1999).

These early projects often commenced without baseline studies. Project success increased to some degree when baseline and vegetation studies were included in the preparation phase; however, a major shortcoming was that projects were trying to work for, but not with, the rangeland users. Training was largely restricted to government officials, and indigenous knowledge was not appreciated. Subsequent social studies and productivity assessments on the basis of the production objectives showed that traditional pastoralists were, in most cases, highly efficient resource managers. It could be shown that there were many sound ecological and economic reasons for pastoral mobility. Pastoralists found themselves forced to act in a certain way, not through ignorance or ill will, but because of difficult legal conditions (i.e. land-tenure systems designed for sedentary agriculture), loss of access to pastures and economic difficulties (low prices for livestock products). Inappropriate interventions were found to lead to poor economic return on investment (Le Gall, 1999). *This triggered a decline of investment in pastoral development on the part of donors and, importantly, also led to a critical revision of concepts and strategies.* Many non-governmental organizations (NGOs) took a very active part in promoting a revision in concepts, and many national governments of developing countries have also advanced considerably in revising their rangeland policies. These trends can be observed within all major donor agencies.

The New Concepts

The critical revision of concepts in rangeland and pastoral development revealed:

- *A high level of efficiency in resource use among many traditional pastoralists.* This is particularly true for sub-Saharan Africa, where it was found that the production of animal protein per hectare in the Sahel was two to three times higher than in areas with similar natural conditions in semi-arid parts of Australia and the USA (Breman and de Wit, 1983). This is also because of the different product mix (meat and wool in the case of ranches in Australia and the USA, whereas African pastoralists keep animals to produce meat, milk, blood, manure and/or draught power).
- *The inadequacy of conventional ecological theory.* Range management is largely based on the theory of succession; that is, removal of grazing pressure allows the vegetation to revert to a climax vegetation. The art of range management – according to this theory – is to stabilize the vegetation at a desired stage. More recent ecological research has shown that in drylands, particularly where annual plants predominate, vegetation yield depends much more on rainfall than on previous grazing pressure. As rainfall varies greatly between years, so does vegetation yield. Under such conditions, the range vegetation is not in equilibrium (Ellis and Swift, 1988; Behnke *et al.*, 1993). In the long run, however, grazing management can influence vegetation composition even in drylands, particularly as far as the balance between woody and herbaceous species is concerned. Management of range can be either 'opportunistic' or 'holistic' or both. Pastoralists practising opportunistic management take advantage of different natural resources when they are available and, when there is little feed and water, try to get by, or move to other pastures. Holistic management looks at resources in a system perspective, and moves and reacts with appropriate management when vegetation is adversely affected by grazing (Le Gall, 1999).
- *A much stronger emphasis on understanding pastoral systems and land use prior to starting development activities.* This refers not only to the ecological reality and to pastoralists' responses to ecological variation, but also to the multiple production objectives of livestock keeping. During the last two decades, a growing awareness of the complexity and wide diversity of pastoral systems and animal husbandry objectives has emerged (cf. de Haan, 1999).

Classifying pastoral production systems according to whether they are oriented more to market or to subsistence and an analysis of price structures of inputs and produce are crucial to understanding differences in supply response to price changes and are therefore a critical element in determining the economic feasibility of investments (de Haan, 1999).

- *A strong move toward decentralization, participatory development and local empowerment* (de Haan, 1999; Le Gall, 1999; Reynolds *et al.*, 1999; Sidahmed, 1999; Turk, 1999). Management of rangelands requires frequent decisions on the part of the users, and the circumstances in which the decisions are made can differ considerably within and between years and seasons. There is now substantial evidence that decentralized and community-based approaches are better suited to natural resource management (NRM) than blueprint-like, centralized ones. There is an almost universal agreement among funding and technical/implementing agencies regarding the need for participatory development. However, there are still substantial difficulties in incorporating a participatory approach into the project-cycle management of the various agencies, since it requires a change in roles of the different partners and in the decision-making processes. Furthermore, different stakeholders may speak different languages, which can make it more difficult to reach agreements between them about resource use. In participatory approaches, local scientists and experts from host countries play a much more important role in project design and implementation (Turk, 1999). The principle of subsidiarity is important in this respect, i.e. what can be decided locally should be dealt with at the local level, and only those issues which require decisions at higher administrative levels should be dealt with there. Care must be taken, however, that higher levels of government assume their responsibilities and that local government, user groups and NGOs are not in charge of tasks which they cannot master (e.g. a local government can declare an area to be affected by a particular disease, but drawing up legal regulations and meeting costs for quarantine are national rather than local tasks).

- *A greater recognition of the multiple functions of rangelands.* Different stakeholders may also have quite different perceptions of the functions and problems of rangelands; for example, as potential carbon sinks (Lusigi and Acquay, 1999); as vast, sparsely populated areas which are locations for water catchment or for plant biodiversity and wildlife refuge; for so-called minor products such as harvesting special mushrooms or resins; as well as pasture for domestic animals.

- *The need for consultation among different groups of range users.* Rangelands are multipurpose areas and there may be different groups competing even for the same use. This is especially true for key resources such as low-lying seasonally inundated areas, where mobile pastoralists and sedentary agro-pastoralists want to graze their animals. Consultation and agreements can create a 'win–win' situation, and such consultative processes can be initiated and supported by projects (Sommerhalter and von Lossau, 1999). However, building up contacts and confidence between different groups can be time-consuming. This approach, therefore, should be attempted only if projects are ensured of a sufficiently long period of funding to give a realistic chance of establishing relations so firmly that they can continue even without project support. As conflicts between different user groups are inevitable, conflict management is an important part of this consultative process. Long-term commitment has already been achieved by some agencies. Although each project phase may be only 2 or 3 years long, the projects of some donors last for 20 years or longer (Turk, 1999). As with other development projects, care should be taken that the procedure for project evaluation and future planning does not interrupt funding, causing temporary cessation of the activities.

- *The need for institutional and human development.* The success of participatory development efforts, and of projects in general, depends to a large extent on the existence of appropriate partners. Including a component of staff training and further qualification of staff in the project design has proved to be very beneficial (Turk, 1999). However, it becomes increasingly clear that appropriate training of producers (e.g. in accounting, or auditing of activities) is crucial for the sustainability of project measures. Official administrative structures in rangeland areas are often poorly developed and, with some justification, regarded as inefficient, while traditional structures have often been eroded and are no longer functioning. Some projects have therefore tried to set up new structures without ensuring links with existing ones. Often, as a result, the new structures no longer have the means to function properly after the end of the project (de Haan, 1999). The reasons for the poor sustainability of project-induced structures should be examined in greater depth so that donors, governments and pastoralists can avoid repeating mistakes and can develop more sustainable institutions for managing the range.

Future Challenges

The challenges for the future faced by international funding agencies concerned with the rangelands will include:

- *Valuing the contribution of rangelands to global ecology and economics.* There are strong indications that demand for food from animal sources will increase strongly over the next few decades (Turk, 1999). It has been estimated that, by 2020, demand for milk and meat will increase annually by 3.3% and 2.8%, respectively. However, the off-take from grazing systems has increased by only 0.4% year^{-1} (de Haan, 1999) and it is doubtful whether this modest increase in production from rangelands can be sustained. On the other hand, rangelands occupy the larger part of the earth's land surface. Le Gall (1999) estimates that one-third of the earth's productive surface are dry rangelands and their ecological role should be better appreciated. Carbon sequestration is only a starting point to valuing the contribution of rangelands to global ecology and economics. Other important issues are maintenance of the diversity of flora and fauna, hydrological cycles and cultural values. This will remain an important challenge on the conceptual level for years to come. It may be more important to safeguard the rangeland's existing productivity and functions rather than to increase animal production.

- *Gaining a better understanding of drought and developing strategies for preparedness.* A key characteristic of rangelands is the spatial and temporal variability of vegetative production and access. Successful pastoralists have developed coping strategies, which may include conservative stocking, trying to develop large enterprises which can survive shocks, seasonal or opportunistic movements of their herds, associations with crop farmers in better-endowed areas, and appropriate livestock marketing strategies. There is some evidence that the apparent increase in drought frequency is not so much a meteorological phenomenon, but is linked to a more intensive utilization of grazing resources and to an erosion of traditional strategies to cope with environmental variability. Although there has been, in recent years, a significant increase in scientific knowledge with respect to early-warning systems before drought, food aid during drought and restocking after drought, government and international donor assistance still includes provision of subsidized or free feed. This can contribute to overstocking and pasture degradation, thus aggravating the effects of droughts as well as increasing their frequencies.

- *Supporting diversification of income and employment.* A diverse portfolio of household income is an important way of reducing the risks in semi-arid and arid rangelands. This is frequently sought through shopkeeping and trade in town, crop farming and investing in real estate in towns and villages. It is an open question whether investment by traders and farmers in the pastoral sphere is desirable or not. It is, after all, one means of diversification.

Supporting diversification is an important element in poverty alleviation. In many situations where animal and land resources are limited, providing support to diversify sources of income can facilitate the exit of part of the population from the pastoral sphere. As many pastoral enterprises operate close to the poverty line, developing alternative sources of income and supporting the people's own efforts in this direction should have high priority. This will also help to reduce dependencies and pressures on rangelands.

The search for alternative occupations and employment should include attention to diversification of rangeland products (resins, medicinal plants, mushrooms) and ecotourism, which is often integrated with wildlife tourism (Reynolds *et al.*, 1999). Possibilities of paying pastoralists for ecological services (maintenance of biodiversity, carbon sequestration) should be explored. For these services, appropriate benefit sharing and monitoring are crucial, and ways should be sought to avoid the corruption that often accompanies subsidies.

- *Better incorporation of pastoralists in the consultative process on rangeland use.* There is a global decline in social cohesion at the higher levels of pastoral organization. Traditional institutions are eroding and pastoral organization is becoming fragmented, while the importance of families and small groups as the main decision-making units is increasing. This, together with the necessity to move, weakens the input of pastoral peoples in consultations regarding natural resource management (NRM). Urban dwellers and crop farmers are often much better represented in these consultations and this bias can accelerate the erosion of pastoral land use. Special efforts are therefore necessary to strengthen the representation of pastoral people in the consultative process. Finding true representatives of pastoral peoples may already be a difficult task. Pastoral associations are assuming an increasingly important role, but often commercial and social problems are confounded by too many NGOs and donors. There is still a long way to go in sensitizing government with respect to the need to support pastoral institutions and empower herders as a cost-effective way of improving NRM.

- *Maintaining adequate access to land, water and key resources.* Increasing human population pressure in rangelands leads to an increased use of higher-potential areas (run-on areas, river valleys, etc.) for cropping. As many crop farmers are also livestock keepers, the areas surrounding croplands are often heavily grazed on a year-round basis. This process can undermine the viability of pastoral systems, especially where it leads to more difficult access to water by mobile herds. Problems of access are exacerbated by land-titling projects, which, by their very nature, favour sedentary over mobile land use. As these higher-potential areas are crucial to pastoralists for coping with seasonal variations and droughts, changing the use of land in a small (but key) area can make the exploitation of significant areas of upland range more difficult, if not impossible. This can reduce livestock production in rangeland areas and, as a result, lower the human-support capacity of the land. Concerted efforts are necessary to preserve adequate access to crucial key resources and to improve the awareness of political decision-makers with respect to the special resource-access needs of mobile pastoral production systems in the dry rangelands. Although there are some reasonably successful projects in this respect (cf. Sommerhalter and von Lossau, 1999), this remains a critical challenge in pastoral development efforts.

- *Ensuring access to key services.* The vast extent of rangelands and the low density of human population pose special problems with regard to provision of services such as water supply, animal and human health care, education or law enforcement. Some services supported by funding agencies rely on para-professionals and user groups working on a private-sector basis. While some such initiatives have produced good results, the sustainability of these systems is still unclear (de Haan, 1999). In countries with high rates of inflation (e.g. Sudan), it is uncertain whether the veterinary medical supplies can be replaced. Staff levels can fluctuate when the local people, once trained, opt to work in urban rather than rural areas. Women have a greater tendency to stay in the rural areas, but household duties and tradition may make it difficult for them to attend training courses outside their immediate home area. If water-points are mechanized (e.g. at deep bores which, in some areas, are the only reliable source of usable water), obtaining spare parts for pumps and tubes for the wells can be a major problem which exceeds the financial capacity of a small community. Provision of adequate education, which is a crucial element to ensure better integration of the pastoral population in the consultative process related to NRM, remains a challenge with respect to both the curriculum and the system of education.

- *Creating a favourable political climate for NRM projects on national and international level.* This is an important prerequisite for successful NRM, but can be influenced only indirectly by the main funding agencies. Nevertheless, there have been some successes. On the international level, the convention for combating desertification, the convention for biodiversity and the Kyoto protocol are some milestones which, although far from being perfect, can give a framework for meaningful NRM programmes. On the other hand, the World Trade Organization agreement has created additional pressures by liberalizing markets. This makes it more difficult to implement payment for ecological services and to maintain some ecological standards in products. Donor agencies and governments in developing countries should work together to use their influence at least to minimize the potentially negative effects of so-called free markets on marketing of rangeland products.

The use of rangelands still provides considerable conceptual difficulties for land-use planners and policy-makers. It exploits natural processes rather than attempting to control them, as is the case in cropping or intensive animal husbandry. It favours flexible and communal land rights rather than private land tenure and it requires some 'fuzzy logic' and site-specific investigation and planning. Although we have already travelled a long way from the top-down transfer-of-technology type of range development project of the 1960s, there is still a great deal of conceptual work to be done. This includes:

- reaching agreement on appropriate and universally accepted definitions and classifications of rangelands;
- gaining a better understanding of the capabilities and limitations of rangeland productivity;
- economic assessment of different range uses, and of complementarities and trade-offs between these uses; and
- development of appropriate institutions for range management and for providing services.

References

Behnke, R.H., Scoones, I. and Kerven, C. (eds) (1993) *Range Ecology at Disequilibrium*. Overseas Development Institute, London.

Breman, H. and de Wit, C.T. (1983) Rangeland productivity in the Sahel. *Science* 221, 1341–1347.

de Haan, C. (1999) Future challenges to international funding in pastoral development. In: Eldridge, D. and Freudenberger, D. (eds) *People of the Rangelands. Building the Future. Proceedings of the VI International Rangeland Congress.* VI International Rangeland Congress Inc., Townsville, Australia, pp. 153–155.

Ellis, J.E. and Swift, D.M. (1988) Stability of African pastoral ecosystems, alternative paradigms and implications for development. *Journal of Range Management* 41(6), 458–459.

Le Gall, F. (1999) World Bank's perspective on rangelands and pastoral development. In: Eldridge, D. and Freudenberger, D. (eds) *People of the Rangelands. Building the Future. Proceedings of the VI International Rangeland Congress.* VI International Rangeland Congress Inc., Townsville, Australia, p. 156.

Lusigi, W.J. and Acquay, H. (1999) The GEF experience in supporting programmes related to land degradation in developing countries. In: Eldridge, D. and Freudenberger, D. (eds) *People of the Rangelands. Building the Future.*

Proceedings of the VI International Rangeland Congress. VI International Rangeland Congress Inc., Townsville, Australia, pp. 156–159.

Niamir-Fuller, M. (1999) International aid for rangeland development: trends and challenges. In: Eldridge, D. and Freudenberger, D. (eds) *People of the Rangelands. Building the Future. Proceedings of the VI International Rangeland Congress.* VI International Rangeland Congress Inc., Townsville, Australia, pp. 147–153.

Reynolds, S., Batello, C. and Baas, S. (1999) Perspectives on range development: the Food and Agriculture Organization of the United Nations (FAO). In: Eldridge, D. and Freudenberger, D. (eds) *People of the Rangelands. Building the Future. Proceedings of the VI International Rangeland Congress.* VI International Rangeland Congress Inc., Townsville, Australia, pp. 160–165.

Sidahmed, A. (1999) The experience of IFAD in supporting rangeland development. In: Eldridge, D. and Freudenberger, D. (eds) *People of the Rangelands. Building the Future. Proceedings of the VI International Rangeland Congress.* VI International Rangeland Congress Inc., Townsville, Australia, p. 159.

Sommerhalter, T. and von Lossau, A. (1999) Lessons learnt and perspectives of GTZ in support of pastoral development in developing countries. In: Eldridge, D. and Freudenberger, D. (eds) *People of the Rangelands. Building the Future. Proceedings of the VI International Rangeland Congress.* VI International Rangeland Congress Inc., Townsville, Australia, pp. 1063–1064.

Turk, J.M. (1999) In defence of the range: USAID's experience in rangelands development projects. In: Eldridge, D. and Freudenberger, D. (eds) *People of the Rangelands. Building the Future. Proceedings of the VI International Rangeland Congress.* VI International Rangeland Congress Inc., Townsville, Australia, pp. 1055–1060.

Policies, Planning and Institutions for Sustainable Resource Use: a Participatory Approach

17

Nick Abel, Mukii Gachugu, Art Langston, David Freudenberger, Mark Howden and Steve Marsden

Introduction

The aim of this chapter is to describe a participatory approach to policy making for rangelands. It was demonstrated in a workshop at the VI International Rangelands Congress (IRC) (Eldridge and Freudenberger, 1999). Our intended audiences are rangeland users, public servants, interest groups and politicians. Our method is one of many possible approaches (IDS, 1998). We begin by discussing the reasons why participation matters in policy-making, before describing our method. Next we discuss some of the proposals for changes to policies advocated by participants in role-plays used during the VI IRC. We conclude by discussing the potential and limitations of our workshop demonstration, and of participatory approaches applied in reality.

Why a Participatory Approach?

Changes to policies and institutions are generally made by politicians and public servants, influenced by pressures from voters, interest groups and providers of resources to political parties (Abel, 1999). The relative importance of these influences varies between countries. Interpretation of the role of stakeholders in this process depends on social theory. To Marxists, who see capitalist society as being fundamentally in conflict, participation is an appeasement of exploited groups. It postpones the structural changes required to resolve conflicts (Dahrendorf, 1959).

Others see society as fundamentally cooperative (Parsons, 1951), and policy-making and institution building as a technical–bureaucratic process in which participation is unnecessary. A third view is that society comprises groups with differing values that pursue their own interests, but form allegiances with other groups with whom they share goals and values – conflict and cooperation both occur (Dahrendorf, 1967; Dale and Bellamy, 1998). Acceptance of the third view is implicit in the approach we are advocating. A framework for understanding the underlying political–economic processes is discussed by Godden (1997).

We contend that opportunities for enhancing the sustainability of rangelands through policy and institutional changes are commonly not recognized or taken. One reason is that politicians and public servants tend to follow Parsons, and behave as if society is fundamentally cooperative. They promote a 'command and control' approach to policy formation (Gunderson *et al.*, 1995). Decision-making is centralized, with little feedback to policy-makers about the social, economic and ecological effects of policy implementation. The lack of feedback has four consequences. First, local knowledge, a rich source of technical and social information, is ignored. Second, potential 'win–win' solutions to sustainability problems – that is, solutions that make all parties to a dispute better off – may go unnoticed. Third, policies and institutions do not adapt to changing local circumstances (Dovers and Dore, 1999). Fourth, even though politicians and public servants may be able to make policy and institutional changes without local participation, local rangeland users are likely to subvert or ignore measures they do not support, making implementation expensive or unfeasible. Thus, although politicians and public servants are usually more powerful than local rangeland users, it is still in their interests to develop policy and institutional proposals in a participatory way. In general, we expect the benefits of participation for a region to exceed the costs.

Having accepted the need for participation, we are faced with another major barrier to communication and understanding. It is the differences in 'mental models' between individuals and groups. Theory holds that a human brain is unable to process and organize incoming information in an unstructured way. Instead it filters and reorganizes information selectively, in accordance with a structured mental model. During the process we tend to shed information that contradicts our mental models, and accept that which confirms it (Mackay, 1994; Abel *et al.*, 1998).

To understand the viewpoint of another person, one needs to understand their mental model, though without necessarily agreeing with it (Kelly, 1955). This chapter discusses a method for improving the mutual understanding of mental models, and for developing proposals for policy and institutional change that potentially integrate differing views. We accept that past conflicts may delay cooperation, sometimes for decades or more, but assert that win–win opportunities often exist and can be identified and implemented eventually. Before describing an approach for identifying such opportunities, we need some definitions.

Some Definitions

We define 'institution' as sets of perennial rules (laws, for example) and the organizations that implement them. These organizations may also implement policies, which are designed to achieve shorter-term goals. We advocate a participatory approach that involves stakeholders in the establishment and modification of policies and institutions to enhance the sustainability of rangeland regions (Abel, 1999). A 'sustainable region' is one where social, economic and ecological systems persist in the long term (Gunderson *et al.*, 1995). A 'participatory approach' is one in which stakeholders are directly involved in the formation and evaluation of proposals for changing policies and institutions, and their involvement affects content.

Method

At the VI IRC, a demonstration was designed to reflect the participatory method, and so was itself participatory. Our method was developed during a participatory research and development project (Abel *et al.*, 1997; Abel, 1999). The demonstration workshop was based on role-play and policy analyses set in two rangeland regions. Both are imaginary and simplified to the level of caricature because we could not address the complexity of actual rangeland regions in the time available. One imaginary region was in a 'developing country', the other in an industrialized one. Both are subject to similar global forces, which were outlined to participants (Box 17.1). Participants

---

Box 17.1.   Global forces affecting rangeland regions.

Global markets have penetrated economies in both countries, and transnational corporations are increasingly powerful. National governments seek capital from overseas to promote economic growth and reduce unemployment. In return transnational corporations, such as the tourism industry, expect freedom to export profits, and the provision of infrastructure and services, otherwise they withhold investments. International conservation groups put pressure on national governments to further their aims for conserving wildlife. They sometimes use international media to do this – governments are sensitive to reports that might influence investments in general, and the tourist industry in particular.

Box 17.2.   Industrialized country.

A democratically elected government depends on electoral support from pastoralists for reliable re-election. However, as the rural population is only 10% of the total and declining, government cannot afford to alienate dominant urban interests, or it will definitely lose elections. One of its priorities is maintenance of regional economies and services in the face of a declining pastoral industry. The decline is caused by increasing costs of production and declining prices for pastoral products. A related priority is the promotion of nature conservation to support a growing tourist industry and satisfy pressure from urban conservationists. More land is therefore needed for national parks.

The tourism industry, conservationists and pastoralists are the stakeholder groups. Tourism is owned mostly by international capital. The industry has a strong influence on government policies through urban and rural investment and job creation. Outside cities and away from beaches, it depends on abundant, tame wildlife and 'wilderness'. National conservationists are city-based. They are able to influence government policies through the media, government being sensitive to the effects a 'bad press' will have on votes and international tourism.

Our pastoral stakeholders live in a region of 100,000 km² where rainfall is low and unreliable; 4% of the region is national park, 5% opportunistic rainfed cropping, the rest is grazed by livestock and wildlife. Pastoral use affects some native plants and animals adversely – the higher the stocking rate, the greater the effect. Some wild herbivores benefit, however, from the provision of water for stock. They compete with livestock during droughts. Pastoralists grow wool and beef on fenced ranches watered from boreholes. They grow crops when and where soil moisture permits. They stock heavily when in debt, and to pay school fees. Stocking density decreases with size of landholding. The pressure to stock heavily increases as prices for pastoral products fall. Pastoral land is owned by the nation, and leased by pastoralists.

Pastoralists need health, banking and other services, and infrastructure. These are declining in parallel with declines in population and regional wealth.

---

were also given more detailed information on the industrialized country (Box 17.2) and the developing country (Box 17.3). Participants were divided into four groups with three sub-groups in each, as shown in Table 17.1.

All participants received a land-use map of the region (not included here). In addition, representatives of each category of sub-group received a briefing note specific to their role. Participants did not see the briefing notes for sub-groups they did not represent. Participants representing one sub-group thus had limited understanding of the 'mental models' of those in other sub-groups. The briefing notes in Box 17.4, the map, and Figs 17.1 and 17.2 were given to participants in sub-groups representing the politician and advisers in the developing country. Sub-groups representing pastoral stakeholders in the developing country were given the briefing note in Box

Box 17.3. Developing country (this information is expanded in Boxes 17.4–17.6).

The government of the developing country is democratically elected. Ethnic allegiance to candidates is important. Government depends on electoral support from pastoralists in the rangeland region for re-election, but it risks losing elections if it alienates the urban electorate by not providing jobs. Policy priorities are related to the fact that the national population is large and growing, distributed at present equally between rural and urban areas, but with migration to towns putting pressure on infrastructure, services and employment.

The rangeland region in the developing country is 10,000 km² and has a sparse and variable rainfall. The interests of its inhabitants are represented by a Member of Parliament. The main stakeholder groups are pastoralists and the tourism industry.

Table 17.1. Division of participants into sub-groups.

Type of country	Group	Sub-groups		
Developing	1	Politician and advisers	Tourism business	Pastoralists
	2	Politician and advisers	Tourism business	Pastoralists
Industrialized	3	Politician and advisers	Conservationists	Pastoralists
	4	Politician and advisers	Conservationists	Pastoralists

Notes:
'Pastoralists' in the industrialized country are also known as ranchers.
1 and 2 are a replicate pair, as are 3 and 4. They were formed to keep the number of participants to a manageable level.

17.5, the map, and Fig. 17.1. Those representing tourist industry stakeholders in the developing country received a map, Fig. 17.2, and the briefing note in Box 17.6. Comparable information was given to groups representing the industrialized country. Participants came from a variety of nations and backgrounds. These were not necessarily reflected in the roles they played.

Each sub-group developed an influence diagram to represent the mental model of the stakeholder group or politician it represented. Those representing pastoralists answered the question: 'What factors affect the welfare of my group?' The tourism industry representatives identified factors that affected the commercial viability of this industry. Conservationists showed the influences affecting the conservation of wild species and ecological communities. Representatives of

Box 17.4.   Politician and advisers in the developing country.

Your Government depends on electoral support from pastoralists in the rangeland region for re-election, but it risks losing elections if it alienates the urban electorate. Policy priorities of government include:

- more land for wildlife, to support the tourism industry;
- beef production, to export for foreign exchange to buy imports for the urban élite, and to feed towns;
- job creation;
- regional economic growth to reduce rural–urban migration;
- increased primary health care;
- better nutrition;
- better education.

You want to do the best for your electorate. After all, you belong to the same ethnic group, but you cannot do anything if you are not re-elected. Without some support from the tourism operators your election campaign will suffer, and one of your rivals has been making big promises to voters. You fear that the tourist operators will withdraw from the region if their needs are not met. You believe it may be possible to satisfy at least some of the requirements of your electors and meet the needs of the tourism industry. Your advisers have supplied some information on the region, summarized in Figs 17.1 and 17.2.

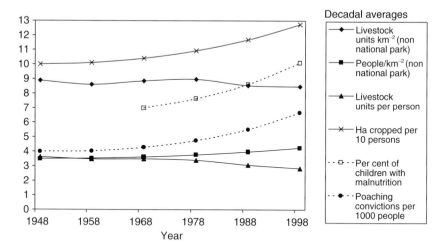

Fig. 17.1.   Regional pastoral trends: developing country.

politicians and advisers answered the question: 'What factors affect my chances of re-election?' Members of the political sub-groups were encouraged to join other sub-groups and learn about the mental models of their stakeholders. Other sub-groups did not mingle while influence diagrams were being constructed.

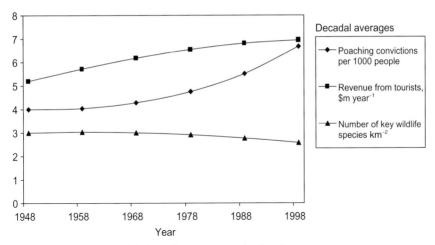

Fig. 17.2. Regional wildlife and tourism trends: developing country.

Box 17.5. Pastoralists in developing country.

You belong to a group of transhumant pastoralists. You are the main ethnic group in a region of 10,000 km². The only other people in your region are some government employees, storekeepers, traders, tourism operators and their clients. Your people are the traditional owners of the land. However, 50 years ago your people were evicted from 3000 km² to create a national park. Grazing is excluded from this under the Wildlife Act. This apart, any member of your ethnic group has access to grazing in the region, subject to seasonal restrictions.

Rainfall is low and unreliable. Water for livestock is accessible over the region from wells dug in riverbeds. Under the institutions of your culture, uplands with lighter soils are grazed during the wet seasons. Grazing on the clay plains is reserved for the dry season. Traditionally, the main source of food for your people has been milk from cattle. Sheep, goats and cattle also provide a little meat. Animals are sold for cash or traded for grain at the one commercial market in the only town. However, animals lose weight during the long walk to market, and prices paid are therefore low, especially during drought, when the price of grain is high. Your people would like to own more livestock, but periodic drought, disease, the prevention of grazing in the national park and the need to sell animals for cash prevents their increase. Some of your animals die from diseases you attribute to wildlife. Large predators from the park also take some of your animals and an occasional person.

As the human population increases, other sources of food are growing in importance. Some crops are grown opportunistically on the plain when the unreliable rainfall permits. You are often in conflict with the regional Soil Conservation Officer over restrictions on ploughing the plain. Despite this, the increase in the number of your people is causing cropland to spread, and there is growing competition between cropping and the provision of dry season grazing. Wild animals often destroy your crops and compete with your livestock at water holes.

Continued

Box 17.5. *Continued*

However, hunting of wildlife is another source of food and cash. Most animals are killed in the national park because they are easier to hunt there, and snares can be left without endangering livestock. From time to time, members of your group are convicted of poaching and jailed. In spite of all the ways your people have adapted, more and more children are being taken ill from lack of food, and there are insufficient clinics to treat them.

Your Member of Parliament was born in this region. He needs your votes if he is to maintain his seat. You often tell him of your many complaints, and expect some action in return for your loyalty.

Box 17.6. Tourism industry in developing country.

You represent an international company with activities in a number of countries. Your main aim is to maintain commercial viability in a competitive industry. You have won and so far kept the exclusive right to bring tourists to this world-famous national park. You contribute to the election funds of the local Member of Parliament. Your clients expect to see calm wild animals at high densities. A few regular visitors have complained about the apparent decline in some species (Fig. 17.2), and this is supported by aerial survey data. Most animals in the park are migratory, and they must pass through land that is not yet designated as national park. Here they catch diseases from livestock, are harassed at water holes and many are killed or wounded. There have been a number of embarrassing incidents lately where animals wounded by poachers have attacked tourist vehicles. You believe the troublesome pastoralists should be moved and the national park extended. The problem is made worse by the reluctance of government to commit funds to tourist roads in the park so more tourists can be brought in. Your company has threatened to withdraw investments from this country if their commercial needs are not met, and go where governments enforce the law rigorously and support economic development.

Figures 17.3–17.5 are influence diagrams from participants in group one (Table 17.1). They represented politicians and advisers, pastoralists and the tourism industry in a developing country. Diagrams were constructed around the question addressed by each group by listing factors affecting it, and grouping and prioritizing those factors.

Figures 17.3–17.5 were copied during the session using the decision support package Vensim (Ventana, 1998). They were slightly edited subsequently to remove repetitions and add clarity. Influence diagrams can be developed in Vensim into quantified simulation models (Walker *et al.*, 1999). That was not appropriate for this exercise.

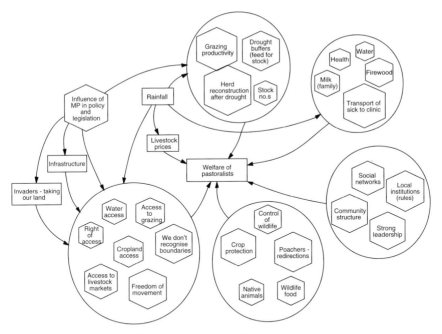

Fig. 17.3.   Mental model of pastoralists: developing country.

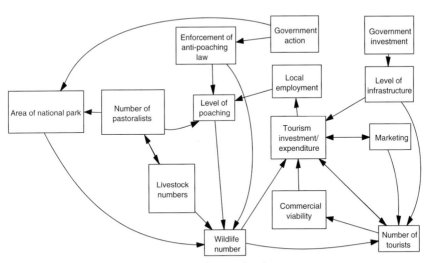

Fig. 17.4.   Mental model of tourism industry: developing country.

Proposals for Changes in Policies and Institutions

Once the mental models of stakeholders and politicians were constructed by the sub-groups, the groups re-formed (Table 17.1) and

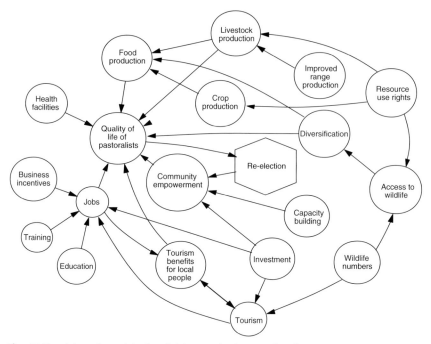

Fig. 17.5. Mental model of politician and advisers: developing country.

sub-groups learned about each others' models. Each group then constructed a table of proposed changes to policies and institutions. Table 17.2 is an example.

The demonstration workshop proved too short to develop these proposals. Sub-groups would in real circumstances be given the opportunity to respond to proposals from other sub-groups and suggest changes. In Table 17.2, column three, we have added contributions to show how we had expected the process to work, with examples of potential win–win solutions identified in bold.

For illustrative purposes we have developed one proposal a little further. The example is the politician's response to proposal 5: establishment of buffer areas around the park where hunting is permitted, but livestock excluded. Elements of proposals 4 (making better use of the local knowledge base), 6 (roads that serve both stakeholder groups) and 7 (a collaborative tourism venture) might be combined in a scheme which brought mutual benefits and resolved some of the resource use conflicts. We would advocate a participatory approach to the development of these preliminary ideas.

Table 17.2. Examples of policy proposals from a developing country group.

Examples of proposed policy and institutional change	Proponent	Response to the proposal by other sub-groups	Respondent
1. Redefine park boundaries to reduce area and give access by livestock to key areas.	P	Not in the interests of regional economic development.	M
		Devastating effects on wildlife, which is a national heritage.	T
2. Give long-term compensation to pastoralists for loss of traditional resource use rights.	P	Good idea. The ex-colonial government and the tourism industry could contribute to a fund.	M
		Good idea. The present government should pay.	T
3. Chief of the pastoralists stands for president.	P	He would not represent interests of the region as well as the incumbent.	M
		He would not represent national economic interests.	T
4. Make better use of our local knowledge base.	P	**Good idea. We could set up a three-way discussion group.**	M
		Good idea. We could teach pastoralists about the evils of poaching.	T
5. Redefine park boundaries to increase the area and take in more areas with good wildlife habitat.	T	We would vote for a different candidate, hunt more and drive our livestock very slowly along tourist roads.	P
		We might instead establish buffer areas around the Park where hunting is permitted, but livestock excluded.	M
6. Extend and improve roads.	T	Good idea, but make sure they connect our settlements to the town and do not go near the National Park.	P
		Perhaps roads can serve both pastoralists and tourism.	M
7. Develop a collaborative tourism venture involving ourselves and the pastoralists. It would include training, jobs and a share of profits, in exchange for ending poaching, keeping	T	We would manage it and allow the tourism industry to participate.	P
		Good idea – please fund a workshop to develop the idea and build support among pastoralists.	M

Continued

Table 17.2. *Continued.*

Examples of proposed policy and institutional change	Proponent	Response to the proposal by other sub-groups	Respondent
livestock out of the Park, and allowing wildlife to migrate freely.			
8. Provide incentives for agricultural production and marketing, and for other local businesses, including tourism.	M	We especially support incentives for tourism. We have just been offered an incentive by a neighbouring country to move our operation over there.	T
		We especially support incentives for keeping cattle.	P
9. Devolve rights over local resources to local people.	M	Not in our interests. We would take up the incentive offered by a neighbouring country to move our operation there.	T
		Good idea – please fund a workshop to develop the idea in consultation with us.	P

Key: M = politician and advisers; P = pastoralists; T = tourism industry; bold text = potential win–win.

Potential and Limitations of the Approach

Next we discuss the potential and limitations of this participatory approach first as a learning tool for policy-makers, then as a method applied in a real region. Key points are shown in italic below.

The role-play approach is a cost-effective way of giving experiences of resource use conflicts and differing perceptions to policy-makers, without inflicting harm on real people and ecosystems. Figures 17.3–17.5 show stark differences between mental models. Of course differences were contrived by providing different information to sub-groups, but the process emphasizes to participants how the same issue can be interpreted in very different ways by protagonists in a resource use dispute. In terms of psychological theory, it provides an experience from which policy-makers are likely to learn more effectively than from, for example, reading or being told about it (Mackay, 1994).

We use a participatory approach to capture and share the mental models of actual stakeholders and policy-makers in a rangeland region in New South Wales, Australia (Abel *et al.*, 1997; Abel, 1999). Our project runs for four years, participants have worked with us from the outset, they are already familiar with the region, and they handle complex information in a series of progressive workshops. The exercise at the VI IRC took 4 hours, participants had not met before, they had no regional knowledge in common, and there was only a single session. The roles played by participants and the circumstances in which they acted were necessarily highly simplified compared with those in a real rangeland region. This could be misleading. We assumed, for example, that the sub-groups are internally homogeneous. Clearly, this is incorrect in reality, given differences in gender, age, wealth and power. Likewise, we simplified societies, economies and ecosystems. If a participatory approach is used to address real policy and institutional issues, time and information needs would be expected to increase with the number of categories of stakeholder, social, economic and ecological complexity, and types and levels of resource use conflicts. We believe that in general the benefits of participation will exceed the costs. *Participation by representatives of stakeholder groups is likely to be a prerequisite for successful implementation of policies and institutional changes in rangeland regions.*

References

Abel, N. (1999) Resilient rangeland regions. In: Eldridge, D. and Freudenberger, D. (eds) *People of the Rangelands. Building the Future. Proceedings of the VI International Rangeland Congress*. VI International Rangeland Congress Inc., Townsville, Australia, pp. 21–30.

Abel, N., Langston, A., Tatnell, B., Ive, J. and Howden, M. (1997) Sustainable use of rangelands in the 21st century: a research and development project in Western New South Wales. *Creating a Green Future*. Australia–New Zealand Ecological Economics Conference. Melbourne, 17–20 November.

Abel, N., Ross, H. and Walker, P. (1998) Mental models in rangeland research, communication and management. *The Rangeland Journal* 20, 77–91.

Dahrendorf, R. (1959) *Class and Class Conflict in Industrial Society*. Routledge and Kegan Paul, London.

Dahrendorf, R. (1967) *Conflict after Class: New Perspectives on the Theory of Social and Political Conflict*. The third Noel Buxton lecture of the University of Essex, 2 March. Longmans, London.

Dale, A. and Bellamy, J. (1998) *Regional Resource Use Planning in Rangelands: an Australian Review*. Occasional Paper 6/98. Land and Water Resources Research and Develoment Corporation, Canberra.

Dover, S. and Dore, J. (1999) Adaptive institutions, organizations and policy processes for river basin and catchment management. Speaker's Paper, pages 123–133 in *2nd International River Management Conference 29 September–2 October, 1999*. Brisbane River Festival Committee.

Eldridge, D. and Freudenberger, D. (eds) (1999) *People of the Rangelands. Building the Future. Proceedings of the VI International Rangeland Congress.* VI International Rangeland Congress Inc., Townsville, Australia.

Godden, D.P. (1997) *Agricultural and Resource Policy: Principles and Practice.* Oxford University Press, Melbourne.

Gunderson, L., Holling, C.S. and Light, S. (1995) *Barriers and Bridges to the Renewal of Ecosystems and Institutions.* Columbia University Press, New York.

IDS (1998) Participatory monitoring and evaluation: learning from change. *Policy Briefing Issue* 12, November. Institute of Development Studies, University of Brighton.

Kelly, G.A. (1955) *The Psychology of Personal Constructs.* Vols I and II. W.W. Norton, New York.

Mackay, H. (1994) *Why Don't People Listen?* Pan, Sydney.

Parsons, T. (1951) *The Social System.* Free Press, Glencoe.

Ventana (1998) *Windows Onto Reality.* Tutorial. Vensim Version 3.0. Ventana Systems, Inc., Harvard.

Walker, P.A., Greiner, R., McDonald, D. and Lyne, V. (1999) The Tourism Futures Simulator: a systems thinking approach. *Environmental Modeling and Software* 14, 59–67.

Economics and Ecology: Working Together for Better Policy

18

Nick Milham

Introduction

Looked at holistically, rangeland is a multiple-use resource. The most cursory perusal of rangelands literature puts this point beyond question. Bartlett *et al.* (1999) and Atkins *et al.* (1999), for example, between them identify more than ten potential commercial and non-commercial uses for rangelands, ranging from grazing of domesticated animals to wood harvesting, mining, tourism and wildlife conservation. There are many more when other aspects of the rangelands, such as water catchment, cultural value and utilization as a pollution buffer/sink, are taken into account.

The behaviour of individual range users around the world, however, suggests that in practice, *most people have a 'unimodal' view of the rangelands*. Rangeland is perceived from the perspective of the principal benefit that the individual expects to derive from it, be it as traditional grazing lands providing food, shelter and security for generations past, present and future, or as a factor of production in a profit-maximizing commercial enterprise, or as a complex natural system supporting a vast diversity of life forms, or as a place to visit for rest and recreation, or as simply the mantle over untold wealth concealed in the geological formations beneath.

It appears that to the extent that other perspectives are appreciated, they are by choice or of necessity subjugated to the pursuit of goals related to the primary interest of the individual (subsistence, exploitation of mineral deposits, wildlife preservation, etc.). There seem to be only few circumstances where persistent bi- or multi-modality may be observed in private behaviour and typically these are a result of the individual being indifferent between alternative uses or where there is a strong degree of complementarity between the alternatives (e.g. ecotourism and the preservation of habitat and wildlife).

Governments, however, are responsible for the long-term welfare of the community as a whole and therefore cannot focus solely on the benefit any one particular group in the community may derive from rangelands. Thus, *a unimodal approach is inadequate for the purposes of government*. Rather, consideration must be given to and compromises made between the entire range of current and possible uses of the resources of the nation, including rangelands. This 'public' perspective of the rangelands is captured succinctly in a recent joint statement by Government Ministers in Australia:

> Australia's rangelands have important ecological significance, are an important economic resource and have significant cultural and heritage values for indigenous and non-indigenous Australians. The management of the rangelands, now and into the future, is therefore of great interest and consequence to the whole Australian community . . .
>
> The challenge is to balance the diverse economic, cultural and social needs of rangeland residents and users with the need to maintain its natural resources and conserve our biological and cultural heritage.
>
> (Commonwealth of Australia, 1999)

The name of virtually any other nation with rangeland ecosystems could be validly substituted for that of Australia in the above statement. *Governments around the world are struggling* in a myriad ways and with varying degrees of success, *to grapple with the challenge of how to manage rangelands to balance the interests of different users so as to deliver the best overall outcome for their respective nations, and the global community*, both now and into the future. Atkins *et al.* (1999), Bartlett *et al.* (1999), McCarthy and Swallow (1999) and Wu and Richard (1999) provide illustrative examples of this struggle.

The difficulty in pursuing a multi-modal approach is that governments do so in the face of conflict between differing interest groups, each attempting to ensure that their principal benefit is maintained. The focus of this chapter is on the potential that is now being realized to bring together the two disciplines of *ecology and economics to provide practical tools to assist governments to develop policies that deal equitably and efficiently with the necessary compromises inherent in rangeland management.*

Government Policy and the Rangelands

In essence, government policy is about intervening in the community to achieve desired change and/or prevent undesired change. It is about encouraging or maintaining (by inducement and/or penalty) certain behaviour by certain citizens in order to achieve an outcome that is considered desirable for the community at large. That is, *government policy is about influencing the way people utilize their resources and the way they relate to each other (within and across state and national boundaries) and their environment.* Thus, in relation to rangelands, governments around the world: regulate domesticated animal stocking rates to reduce the likelihood of range degradation; subsidize the establishment of human and animal watering points to broaden grazing opportunities and support local communities; provide taxation incentives to pastoralists to encourage early destocking at the onset of drought; regulate land development to protect native flora and fauna and biodiversity; require landholders to control pests and diseases, feral animals and weeds; provide financial assistance to farmers to sustain their families and businesses during temporary industry downturns; and change land tenure arrangements for a gamut of reasons.

Overlaying these interventions is the general policy environment of taxation arrangements, interest rate and exchange rate management, social welfare measures, industry and regional development programmes, social reform, defence, international aid, environment protection and wildlife and biodiversity conservation, food security, protection of animal welfare, international trade, etc. While the inhabitants and managers of rangelands therefore have a large number of government policies and programmes seeking to influence their behaviour and its impact on rangeland, governments around the world tend not to have well defined 'rangeland' policies. This is understandable, however, because the interest of government is rarely in any particular natural resource *per se*, but in the current and future contribution of that resource to human society. Hence 'rangeland' policy commonly reflects a wide array of more general government policy objectives.

These policy objectives are not always in accord, and their relative priority may be different in different countries and regions within countries and may vary over time. In addition, their potential impact on rangeland managers may have initially been poorly estimated or even totally overlooked. It is not surprising therefore that examples of ineffective policy settings and even perverse programme outcomes can be observed in rangeland regions around the world. Examples of economic policies which inadequately account for environmental or social impacts, and conservation policies which do not adequately ameliorate other private decision drivers (such as survival), are easy to find.

As an example, in Australia, where government policy at all levels is to encourage and assist early destocking at the onset of drought, a number of programmes put in place to achieve this outcome have had directly contrary affects on pastoral activity. One such programme involved subsidies on freight costs for: (i) the transport of fodder and water into droughted areas; and (ii) the transport of livestock away from droughted areas and their return after the drought had broken. The fodder and water subsidies assisted in keeping stock alive during the drought, but had the very undesirable effect of maintaining stock numbers, and thereby grazing pressure, on the range during a vulnerable period. Moreover, the transport subsidies were not triggered until a 'dry spell' was officially declared to be a drought, and it was found that they provided an incentive for pastoralists to hold stock pending availability of the subsidies, rather than destocking early (Synapse Consulting, 1992; Worrell *et al.*, 1998).

Similarly, as observed by McCarthy and Swallow (1999), while governments recognize the benefits (sometimes necessity) of range managers and inhabitants seeking cooperative solutions, many government programmes have an objective of risk reduction of one form or another. Sharing risk is a natural inducement to cooperate and work collectively; reducing risk weakens this incentive and therefore puts pressure on the administering institution to identify or 'manufacture' other incentives for cooperation.

Many commentators (e.g. Lunney *et al.*, 1997; Wu and Richard, 1999) have highlighted the difficulty in formulating appropriate rangeland policy. The 'barb in the tail' of this problem is that, appropriate or not, once a government policy is put in place it may be very difficult to change or reverse. *This chapter puts forward a vision for economists, ecologists, range administrators and range users to implement a collaborative approach. The challenge is to think about the way the disciplines of economics and ecology, which have traditionally been seen to be in conflict, can be used in concert to value-add to the information available to policy-makers, thereby contributing to the development of improved rangelands policy.* As discussed in the following sections, while substantial challenges remain, there is a heartening degree of enthusiasm within these disciplinary areas and the policy arena for cross-disciplinary approaches to policy advice and development.

Informed Policy Formulation

It is pertinent to make three general observations on the process of government policy development:

1. While political or other perceived imperatives sometimes dictate policy formation in a virtual knowledge vacuum, the more usual approach is for a policy proposal to receive some level of critical evaluation before it is implemented. Thus, *policy-makers need information and expert advice to support their work, and the information/advice they need may be drawn from a broad spectrum of sources and disciplines.* At this level, discipline-based advice may have very limited value: the most useful advice will be that which draws together information from all relevant fields into a coherent framework.

2. Policy-makers commonly ask the question: 'If we do such and such, what will happen?' That is, *they want information/advice presented to them which has some capacity to predict the likely outcome of proposed, and alternative, policy settings.*

3. Economic objectives of governments frequently take primacy over ecological objectives, and ecological objectives are commonly established within economic constraints (see, for example, Lynam and Dangerfield, 1999; Pannell, 1999). Government intervention is never costless and no government has an unlimited budget. Moreover, the costs and benefits of such intervention are commonly distributed differently throughout the community and over time, and government members and officials are always conscious of these distributional effects and the relative balance of the community approbation and ire they may subsequently experience.

Implications for Range Ecologists and Economists

Although derived from a common linguistic root, ecology and economics are widely varying and often conflicting disciplines, each with temporal and spatial frameworks that seem to preclude the concerns of the other. This conflict becomes an important consideration in environmental management for which economic and political time frames and spatial boundaries are usually too short and too small to accommodate the dimensions of entire ecosystems and evolutionary time.

(Lunney *et al.*, 1997)

There is no doubt that ecologists and economists look at the world through different eyes, and the evidence of history is that these differing views have frequently led to conflict between the professions, with both sides on occasion belittling the other. The truth of the matter is that neither discipline, as traditionally practised, can claim to provide in all circumstances the perfect foundation for rangeland policy development. Rather, *both have knowledge and insights to offer and the development of sound rangeland policy depends on contributions from both fields.*

Recognition of this has seen an increasing spirit of cooperation between practitioners and, as part of the more general push for multidisciplinary approaches, the development of the quasi-discipline of 'ecological economics'. From a 'public' perspective, both ecology and economics have uses and limitations in negotiating compromise situations. Ecological economics seeks to overcome at least some of these limitations by utilizing the two disciplines in concert.

> Ecological economics addresses the relationships between ecosystems and economic systems in the broadest sense. These relationships are the locus of many of our most pressing current problems (i.e., sustainability, acid rain, global warming, species extinction, wealth distribution), but they are not well covered by any existing discipline. Environmental and resource economics, as it is currently practiced, covers only the application of neo-classical economics to environmental and resource problems. Ecology, as it is currently practiced, sometimes deals with human impacts on ecosystems, but the more common tendency is to stick to 'natural' systems. Ecological economics aims to extend these modest areas of overlap.
>
> (Costanza, 1989)

These ideals were expressed more than 10 years ago, but how far have we actually come since then? The evidence suggests that, perhaps as a function of necessity, reasonable progress has been made. A brief review of some recent literature illustrates the diversity of possibilities and advantages in a multidisciplinary approach to exploring issues such as the interplay between private decisions on range utilization, the state of range ecosystems, and government policy settings. (In this regard, it is relevant to note that ecological economics and eco-ecological modelling techniques are a component part of the even more holistic, complex adaptive systems approach; see, for example, Abel, 1999.) For example:

- Perrings and Walker (1999) used a sophisticated economic model to examine the relative importance of factors influencing private decisions to conserve native flora and fauna in semi-arid savannas in central and southern Africa. The model, which derives from an economic paradigm but incorporates ecological variables and relationships, is predictive and clearly shows how government policy can influence the relative 'values' of the private incentives to either conserve natural flora and fauna or to graze more heavily.
- Nature conservation has, in fact, been a major practical application of eco-ecological modelling, with writers such as Lehane (1999), who describes an application in relation to elephants in China, and Alexander and Shields (1999) and Lunney et al. (1997) further exploring its practical application in conservation policy.

- McCarthy and Swallow (1999) explored the impact of spatial and temporal production risk on property rights and consequently the effectiveness of range management policies and administering institutions in sub-Saharan Africa; and
- Atkins *et al.* (1999) and Bartlett *et al.* (1999) each described quite different practical approaches used by government (in Western Australia and the US, respectively) to identify and attempt to manage the demands for multiple use of rangeland resources. These approaches, including those used by others, such as Cameron (1997), have sought to combine the insights into human and institutional behaviour provided by a number of disciplines, to lead to better policy outcomes.

Many would agree, however, that there is still an underlying feeling that, to some extent, cross-disciplinary cooperation between economists and ecologists is happening under sufferance. This tension can be a problem if it is allowed to interfere with progress, but it can also be put to considerable advantage. The challenge of responding positively to constructive criticism prevents unthinking acceptance of the mores of your own discipline and encourages professional growth and development. The innovative thinking underlying models such as those developed by Perrings and Walker (1999) and McCarthy and Swallow (1999) demonstrates the benefits to be gained by both professions from multidisciplinary cooperation. These benefits then flow on to policy-makers through better knowledge and advice. Equally importantly, these two papers demonstrate the power of predictive multidisciplinary approaches in: (i) providing a framework for identifying where individual professions fit into the bigger picture; and, by capturing competing objectives, (ii) indicating priority areas for targeting economic and scientific research and policy reform effort, which intuitively has strong potential to influence policy-makers by translating policy proposals into likely outcomes.

In summary, *one of the principal challenges for rangeland professionals is to maintain perspective, to keep an eye on the 'big picture'* and where personal endeavours and individual disciplines fit into it, and not to be totally immersed in the detail of the specific task at hand. This is not a trivial issue: even with the best of intentions, the challenge of being non-reductionist is a difficult one, not least because of institutional factors such as professional training and reward systems that tend to be discipline-focused and favour ever narrowing specialization. It must be appreciated, however, that in the development of rangeland policy, technical excellence at the disciplinary level is the means to an end (securing the long-term future of the rangelands and rangeland communities) not an end in itself.

A further challenge, particularly to the research community, is to (as far as possible) *design projects so as to generate data which provide insight into predictive relationships.* As noted above, understandably enough, policy-makers and administrators have considerable interest in the 'What ifs?' of alternative policy settings.

Challenges For Policy-makers

The foregoing observations highlight a number of challenges for government policy-makers, policy advisers and the administrators of government programmes impacting on rangelands and range communities. The first of these is to recognize and appreciate that *management of rangelands inherently involves compromise between competing uses and objectives, none of which necessarily have any pre-ordained right to dominate.* To enable the merits of alternative policy proposals to be objectively evaluated (both *ex ante* and *ex post*), government policy objectives should be explicit and transparent.

Second, it is essential that the *makers of rangeland policy be alert to the broader policy environment and how it impacts on the rangeland managers* that they are trying to influence. Regardless of the basis or purpose for which rangeland policies (and other policies which impact on rangelands) are formulated, they often have multiple impacts: ecological, economic and social.

These two points emphasize the need to ensure that rangeland policy is well-informed and soundly based on knowledge and practical experience of rangeland environments and the communities within them. The search and assessment process in policy formulation therefore requires appreciation of all sources of information and all points of view. There is also an obvious need for policy-makers and administrators to avoid the trap of relying on convention, to be flexible and, where appropriate and politically feasible, prepared to try innovative approaches to policy and procedures. *The challenge of building a sustainable future for rangelands requires policy-makers, advisers and range professionals to be lateral thinkers, receptive to new ideas.*

The final message is a general warning to government to be prudent and cautious in establishing rangeland policy. There is much yet to be learned about rangelands and the consequences of even apparently minor policy decisions can be pervasive and long-lasting, and may be very difficult – politically, socially and environmentally – to reverse.

Concluding Comment on Implications for Range Users

The emphasis of this chapter is at the level of government policy development and professional input to that process. There are, however, a number of obvious implications for range users. Key points in this regard are that particular groups of range users, e.g. pastoralists, conservationists, miners, etc.:

- must be aware of the policy-making process and have a degree of familiarity with the relevant parts of government (institutions, Ministers, etc.);
- should ensure these parts of government are well informed on their interests so that those interests can be given due consideration in the policy development process; and
- should be aware of and be able to advise policy-makers on conflicts, complementarities and possible compromises between their preferred use and potential competing uses of the range.

Disclaimer

The views expressed in this chapter are those of the author and do not necessarily reflect the policy of NSW Agriculture or the NSW Government.

References

Abel, N. (1999) Resilient rangeland regions. In: Eldridge, D. and Freudenberger, D. (eds) *People of the Rangelands. Building the Future. Proceedings of the VI International Rangeland Congress*. VI International Rangeland Congress Inc., Townsville, Australia, pp. 21–30.

Alexander, R.R. and Shields, D.W. (1999) *Bioeconomic Modelling of Endangered Species Conservation*. Discussion Paper in Natural Resource and Environmental Economics No.19, Centre for Applied Economics and Policy Studies, Department of Applied and International Economics, Massey University, Palmerston North, New Zealand.

Atkins, D., Hunt, L.P., Holm, A. McR., Burnside, D.G. and Fitzgerald, D.R. (1999) Land-use values in the goldfields of Western Australia and their use in regional resource use planning. In: Eldridge, D. and Freudenberger, D. (eds) *People of the Rangelands. Building the Future. Proceedings of the VI International Rangeland Congress*. VI International Rangeland Congress Inc., Townsville, Australia, pp. 1010–1011.

Bartlett, E.T., Van Tassell, L.W. and Mitchell, J.E. (1999) The future use of grazed forages in the United States. In: Eldridge, D. and Freudenberger, D. (eds) *People of the Rangelands. Building the Future. Proceedings of the VI International*

Rangeland Congress. VI International Rangeland Congress Inc., Townsville, Australia, pp. 1008–1009.

Cameron, J.I. (1997) Applying socio-ecological economics: a case study of contingent valuation and integrated catchment management. *Ecological Economics* 23, 155–165.

Commonwealth of Australia (1999) *National Principles and Guidelines for Rangeland Management.* Australian and New Zealand Environment and Conservation Council (ANZECC) and Agriculture and Resource Management Council of Australia and New Zealand (ARMCANZ), Canberra, Australia.

Costanza, R. (1989) What is ecological economics? *Ecological Economics* 1, 1–9.

Lehane, R. (1999) Ecological economics – balancing conservation and development. *Partners in Research for Development* 12, 47–52.

Lunney, D., Pressey, R., Archer, M., Hand, S., Godthelp, H. and Curtin, A. (1997) Integrating ecology and economics: illustrating the need to resolve the conflicts of space and time. *Ecological Economics* 23, 135–143.

Lynam, T. and Dangerfield, M. (1999) Optimal biodiversity conservation in rangelands. In: Eldridge, D. and Freudenberger, D. (eds) *People of the Rangelands. Building the Future. Proceedings of the VI International Rangeland Congress.* VI International Rangeland Congress Inc., Townsville, Australia, pp. 825–831.

McCarthy, N. and Swallow, B. (1999) Managing the natural capital of African rangelands: rights, risk and responses. In: Eldridge, D. and Freudenberger, D. (eds) *People of the Rangelands. Building the Future. Proceedings of the VI International Rangeland Congress.* VI International Rangeland Congress Inc., Townsville, Australia, pp. 1002–1007.

Pannell, D.J. (1999) Paying for weeds: economics and policies for weeds in extensive land systems. In: Eldridge, D. and Freudenberger, D. (eds) *People of the Rangelands. Building the Future. Proceedings of the VI International Rangeland Congress.* VI International Rangeland Congress Inc., Townsville, Australia, pp. 571–576.

Perrings, C. and Walker, B. (1999) Optimal biodiversity conservation in rangelands. In: Eldridge, D. and Freudenberger, D. (eds) *People of the Rangelands. Building the Future. Proceedings of the VI International Rangeland Congress.* VI International Rangeland Congress Inc., Townsville, Australia, pp. 993–1002.

Synapse Consulting (1992) *Report of the Review of the Rural Adjustment Scheme.* Synapse Consulting Pty Ltd, Brisbane, Australia.

Worrell, M., Milham, N. and Curthoys, C. (1998) Drought relief and rural adjustment policy in Australia. Contributed paper to the FAO/ODI 'Drought Live' Electronic Conference.

Wu, N. and Richard, C. (1999) The privatization process of rangeland and its impact on the pastoral dynamics in the Hindu-Kush Himalaya: the case of western Sichuan, China. In: Eldridge, D. and Freudenberger, D. (eds) *People of the Rangelands. Building the Future. Proceedings of the VI International Rangeland Congress.* VI International Rangeland Congress Inc., Townsville, Australia, pp. 14–21.

Building the Future: Practical Challenges

19

Joe Kotsokoane

The Aims of Development

Development is about people, not things. Whether we are academics, bureaucrats or ordinary workers, the purpose of our development efforts is, in fact, to empower people. We want to give individuals and communities the capacity and capabilities to take charge of their own lives; that is, to become self-reliant. Unless we achieve that objective we are in danger of failing to achieve the real purpose of development. This is a big problem in Africa and in many other parts of the developing world. People lack the skills, and the knowledge acquired through formal and non-formal education, to be able to take responsibility for themselves, to sustainably exploit the resource base and to understand the effects and complexity of climate. The international community, under the auspices of numerous development agencies, has failed to eliminate poverty in the so-called Third World. Fifty years ago the purpose of development was to eliminate poverty. However, poverty around the world is now worse than it has ever been before. Many developing countries are worse off than they were half a century ago and the reason is not hard to find. It is the failure to develop capacity, capability and self-reliance. *Too much aid has been directed to short-term economic objectives (which are seldom achieved) instead of long-term human resource development.* Our focus should be on human resource development.

African Agriculture

We seldom ask how we as agriculturalists fit into the development scenario. At a recent seminar in Johannesburg, organized by FAO and the Netherlands government and supported by the International Fund for Agricultural Development (IFAD), we asked what, in fact, is agriculture and what are its functions. *Agriculture, of which rangeland science is a discipline, operates in a complex environment, encompassing political, economic and social issues.* Considering the numerous linkages of this sector, from production through to utilization, its contribution to gross domestic product (GDP) is often underestimated. In any case, inadequate or poor records in developing countries seldom reflect the true state of affairs. In South Africa, economists usually say agriculture contributes only about 5% to the GDP; that is, not nearly as much as manufacturing or mining. They forget that, particularly in developing countries, there is a lot of information about agriculture that is not included in the calculation of GDP. How does one collect and collate reliable statistics about all the activities associated with agriculture? According to IFAD, these include food security, policies and institutions, economic development, poverty reduction in equity, social cohesion, environmental restoration protection and enhancement, science, technology and knowledge, and management of lending sources. How does one take all these into account in calculating the GDP? In a nutshell, agriculture has multiple functions that require a multiplicity of individual and collective skills. Its human base is a spectrum of producers in Africa ranging from the destitute to the affluent. We have what you might call a pyramid of producers in Africa, from the person with one acre of land who is the real subsistence farmer, to people who are part-time farmers and employed by government or who are lawyers or teachers and farm on a part-time basis. A small group at the apex of the pyramid are the emerging true farmers with knowledge, skills, interests and even a few assets to invest in agriculture. Understanding this structure is important because, at the moment, many African governments including South Africa are asking themselves which levels they should focus on. As an example, my own Minister for Agriculture said: 'I haven't got the resources to help everybody but I'd like it at the end of five years after my tenure of office, to establish two hundred or three hundred black commercial farmers'. We still have the problem of trying to empower people who were marginalized or excluded from development by apartheid. But this is not a problem that is peculiar to South Africa; in Zimbabwe, Zambia, Mozambique, for instance, you find that there is no smallholder sector, and development workers and donor agencies often wonder where to start and how long to continue, that is, when to come in and when to exit. These are some of the issues that many people apparently do not comprehend. We are not dealing with commercial farmers but a spectrum of people at various stages of development. *The challenge facing African governments is how to categorize, assist and provide services to the different categories in accordance with their needs.* In the Biblical story of the sower who went out to sow, some seeds fell on good ground, some on fragile ground, some on rocks, some were eaten by birds. A similar procedure has been followed over the past 50 years in trying to bring about African development. It has not worked: we have little food security or income generation. How should we alter our focus to make an impact, and get people to take on commercial agricultural training, both domestic and international?

Role of Government

While those in developed countries are worried about international competitiveness, the developing world is worried about production. The two are at different ends of the development scale. The developing world is at the beginning of the cycle and one might ask why it is trying to do some of the things that the developed world has done, when the developed world is itself beginning to question whether what it has done was right. And yet the developing world blindly follows. We say to the World Trade Organization: 'This is what we want to do', but in fact we see a situation in the developed world where people are taking stock of past mistakes.

In developing countries it is difficult to bring issues to the ears of the authorities. They do not quite understand the importance of exposing policy-makers to the complexities of agriculture and rangeland.

The role of government is to create an enabling environment through appropriate policies, legislation, structures, systems and procedures. *Over and above the effective and timely delivery of services, good governance implies transparency and accountability.* Government institutions should be staffed by knowledgeable, competent and honest public servants with a strong sense of commitment and dedication. This may seem obvious but we have still got to make sure that the government we elect knows what to do, and the people we appoint as civil servants have the ability, the knowledge and integrity to do the things that we expect of them.

The developed world cannot help the developing world unless it has an understanding of the problems. In any context, security of persons and property is paramount. Conflict, violence and instability are inimical to development. Government intervention should be limited to protection of the national interest while making sure that people and property are safe and secure.

Donors and Project Implementation

In the African context, the role of chieftainship, which is associated with land rights and ethnic identity, needs to be redefined in line with changing circumstances. Progress has been retarded and many lives have been lost in conflicts over grazing rights within and between tribal areas. I emphasize the importance of stability of good governments without which there can be no development. The activities of donors require clear definition and monitoring to ensure that they fall within the parameters of government policy and do not, as often happens, lead to unauthorized expenditure by government departments. Conflict often arises when government is unable to meet additional costs resulting from donor-funded projects. *Sometimes donors come with their own agendas, rather than work to the agenda of the government.* If and when they do conform to the government's agenda, they do things that entail additional expenditure over what has been allocated by government. For instance, donors may start a school, put up a building, buy equipment, and even appoint a teacher. When they withdraw their aid and ask the government to take over, they find that government has not budgeted for those expenses. These problems are still occurring. In South Africa we have set up the National Development Agency which explains to donors that we don't want to run them, we don't want to control them, but *we want to make sure that they are going to work within the parameters of government* and that they must not cause us to incur additional expenditure.

Educational and Technological Constraints

The state of education and communication in the developing world imposes severe constraints on development. I do not think we will ever catch up with those who are using computer models, those who do not understand that when I want their address they should not give me their e-mail; they do not realize I do not even have a telephone or a fax! That is the real world in which we live. I work on the Johannesburg fresh produce market helping my farmers to sell their produce. Their biggest problem is getting market information. The people at the market want to pass on information, the farmers want it but there is no communication, no telephones and no faxes. These are the practical realities on the ground; the constraints that freeze people in the developing world.

People in the developing world require good basic education. They must become literate, and be able to keep records. In many situations there are no arrangements for keeping records. Even simple things like birth certificates and death certificates are not available. This is a very serious problem. An illiterate person cannot use an ATM (automatic teller machine). These are the practical realities of the need for human resource development at the lowest levels. The developing world needs basic literacy so as to increase performance and receptivity of new ideas. Formal education needs serious attention, particularly technical and vocational education. It is very easy for the developed world to say Africa is backward and that although they have been working in Africa for a long time the people have not developed yet. The kind of education they made available was academic to the point that we became dependent on others doing the things that needed to be done. We did not have the technical and managerial skills that were necessary to undertake development. Formal education should impart not only technical and vocational skills, but also human values. In many parts of Africa, so-called development has disrupted family and community life. Particularly in areas where there have been wars of liberation, many children have been uprooted. We have a generation of children who have not gone through school, who want employment but have no skills of any kind. Education is one of the spheres where we need to take action and inculcate positive values as well as practical and vocational skills. There are parts of the world where there are surpluses of technical manpower, but much of the developing world is a long way from that situation. We are now looking at our curricula, our facilities and our manpower to try and make them relevant to the development process. This is also true of many parts of the world where there is a lack of skills to cope with modern development.

Multidisciplinary Cooperation

An issue which concerns me greatly is that after 50 years we are still only talking about multidisciplinary cooperation. This problem should have been solved by now; those involved should be working cooperatively and understanding the need for cooperation and collaboration, but alas it is not happening. *People still have their little empires and rivalries that make it difficult to achieve the agreed objectives.* In South Africa, the Farming Systems people, the Grasslands Society, the Economics Association, and the Irrigation Engineers, etc. still have not developed the idea of a common approach. They talk about it, but rivalry is not conducive to development and it creates problems on the ground. Can you imagine the Department of Agriculture calling a meeting to teach people about producing good food? The next day the Department of Education calls the same people to talk about good nutrition. On the third day the Department of Health tells them about the value of food. Many members of the Agriculture Research Council are not sufficiently aware of the practical problems that are created by rivalries and lack of cooperation.

Rural Extension

With regard to extension and information transfer there is a problem of limited resources, in terms of both manpower and facilities, that inhibits the practical application of science and technology. Research and extension capacity in terms of both competence and technical knowledge have a long way to go because many extension workers cling to outdated ideas and still have to change their approach and attitude to meet the needs of new developments. In eastern and southern Africa, research and extension linkages are being strengthened through the Farming Systems approach which also accommodates indigenous knowledge systems including the use of animal traction for cultivation and transport. For example, a conference addressing eastern and southern Africa, supported by the universities of Reading and Edinburgh, is looking at the problems of animal traction as an alternative technology for those people who cannot afford tractors, or who, because they are farming small areas, should not invest in tractors.

On the environmental front we need to create an awareness of the interrelationships of land, water and vegetation as well as the effects of climate and human activity on the resource base. *There is some indigenous knowledge on the ground but we need to make people more aware of the consequences of not protecting the resource base.* There is a need to look at the degradation of grasslands and other ecosystems. In most parts of the world, including southern Africa, past efforts in regard to soil and water conservation have been merely about engineering, building dams and contour furrows but completely ignoring the biological component. Even more importantly, the people were not educated to appreciate their dependence on the resource base through the careful management of resources.

Social Issues

There are a number of social, cultural and gender issues that impinge on development. For example, people talk light-heartedly about the laziness of African men who sit around in the shade while African women are hard at work. That is a completely false impression because there is a division of labour which determines gender roles. Cultural changes are happening in African societies and it is important to recognize that they should not be forced. It is of concern that forces outside the developing countries suggest that their gender problems are also the problems of African women. We still have ministers of government who are polygamists. How do you tell a minister of government who is a polygamist that polygamy is wrong?

Two major issues threatening agriculture are sharp increases in overall population and an ageing farm population. Many young people go to university and want to get into government, but very few of them want to go into farming. We need to make farming more attractive to young people so they can replace the ageing farm population, and we need to pressure the politicians to talk about population control because this is a basic problem requiring urgent action even if it is fraught with social and political dangers.

Conclusion

I have attempted to give an indication of the sort of practical problems that are facing the developing world. They must be dealt with before we can really worry about luxuries such as computer modelling. Technical developments are good but the developing world is still at a stage where *we are saying come and look at what we are doing, join us where we are and help us to move forward*. In my view we will never really make solid progress as long as we allow ourselves to be judged by a yardstick that is inappropriate. For a long time we will find ourselves simply following behind the people who are trying to draw us forward. *We've got to find ourselves, find our base and try to develop ourselves from where we are, and only in that way do I think we will make real positive progress*.

Rangeland Livelihoods in the 21st Century

20

Brian Walker

Introduction

A thorough review of the VI International Rangeland Congress would require careful consideration of the 483 papers, from 77 countries, that were published in the Congress Proceedings. My perspective is that of a scientist. What struck me, as I am sure it did many others, is the consolidation of a trend that emerged at the preceding Congress in Utah in 1995. At that Congress I demonstrated the decline over the last 100 years in the contribution that livestock production makes to the Australian gross domestic product (GDP). Since then, the trend has continued and increased; agriculture as a whole has just about maintained its contribution, but wool prices and other factors have worsened the position for the rangelands. Rangeland livestock production now contributes probably less than about 0.3% of GDP. The same picture holds for most other rangeland regions. Tom Bartlett's scenario (Bartlett *et al.*, 1999) and Barney Foran's nine global drivers (Foran and Howden, 1999) support this proposition.

Also at the Utah Congress, socio-economics emerged as a major gap in our understanding and research. My summary of contemporary papers in rangeland journals at that time revealed that only 13% of the papers could be described as socio-economics, and that most of these were of a local nature (Walker, 1995). A quick categorization of the papers presented at the VI International Rangeland Congress shows that some 23% of them fall into the socio-economics area. This is still a

©CAB *International* 2002. *Global Rangelands: Progress and Prospects*
(eds A.C. Grice and K.C. Hodgkinson)

small percentage, but nevertheless an increase of 10%, and almost a doubling from 4 years ago. This indicates progress.

My reading of these papers indicates that a dominant issue in the rangelands today is sustainable habitation – how to maintain healthy, viable human societies in rangeland regions. Clearly, there is no one solution to this challenge; no magic bullet or technological fix, and no one economic solution. Many different solutions were proposed at this the VI International Rangeland Congress and these solutions, or suggestions for the way forward, fall under the following headings:

1. Better technology. These include increased efficiency of livestock production, pest and disease control, better water management, improved animal genetics and new molecular techniques. There are still advances to be made in the technological area that will improve the benefit : cost ratios for livestock producers.

2. Multiple use. It is already a fact that the rangelands are no longer used only for livestock production, but using the icon of pastoralism for tourism (free-range, clean and green products compared to unhappy, hormone-full animals in feedlots) is a plus for the marketing of the rangeland products. In addition, use of native biota and niche-market crops contributes to income under multiple use. Atkins *et al.* (1999) gave us the value of using road gravel and speciality timbers. There is a long way to go in developing multiple use to its full potential.

3. Pastoral nomadism. Nomadism has been traditionally practised on four continents – Australia, Africa, Asia and America (the USA). To be effective it must operate at an appropriately large scale. In the early development of the rangelands in Australia pastoral nomadism was practised by the famous Kidman enterprise. The use of summer and winter grazing in the USA, the seasonal movement of livestock from the low veldt to the high veldt in southern Africa, also developed in response to the recognized need to use spatial variation in the rangelands.

4. Other values. The thoughtful paper by Guy Fitzhardinge (Fitzhardinge, 1999) emphasized the need to consider other values that society places on the rangelands. We need a systematic, qualitative cataloguing of the sources and the consumers of rangeland services. The services provide benefits locally, regionally, nationally and globally.

5. Participatory planning. Excellent examples of the need for participatory planning, and the benefits it provides in achieving regional goals, were described in the regional scale projects of Nick Abel and his colleagues, by Simon Campbell, and by the desert uplands project.

6. Stewardship. I note here the paper by Mitchell and co-authors (Mitchell *et al.*, 1999) on the applicability of the Montreal process indicators, derived to ensure sustainable harvesting of forests, to rangelands. This process is a sleeper. One can argue the effects of the Montreal process both ways, depending on the actual standards that are applied.

7. Institutional arrangements. Lusigi and Acquay (1999) make a strong statement that, for rangelands in Africa, what is needed is an enabling environment. This is taken to mean security of tenure, dissemination of innovation, and so forth. Christo Fabricius (1999) points out that success depends on institutions and enabling mechanisms. Nancy McCarthy and Swallow (1999) introduced the importance of fuzzy access rights in institutional arrangements.

All of the solutions are valid – and all of them are partial. None of them, on their own, will work as *the* solution to achieving sustainable habitation in the rangelands. A major shortcoming of research in resource management all over the world has been the quest for some particular solution – the magic bullet of a techno-fix, getting the economics right (market-based instruments), or the currently-popular participatory planning, and so on. On their own they will all fail because the answer to sustainability in the rangelands, as in all regional-scale resource use systems, lies in an integrated approach that addresses the combined biophysical, social and economic system.

To emphasize this point I conclude my overview of the important processes by noting four other papers: Tim O'Connor (1999) highlighted the real difficulty, perhaps the futility, of finding general principles; Bob Scholes (1999) found specific science has not had much benefit; Tim Lynam and Mark Dangerfield (1999) were most damning in their assessment of rangeland science. According to them science has basically failed in terms of improving the lot of rangelands and of the people in them. They conclude their assessment by pointing to the possibility of a complex adaptive system approach, and François Bousquet *et al.* (1999) likewise showed how such an approach might work. Complex adaptive rangelands, these authors assert, require complex adaptive models. I take my cue from them. A complex adaptive systems approach requires one to address each of the three sub-systems (ecological, economic, social) simultaneously. I will conclude therefore, by presenting a brief account of some recent work on an integrated model of rangelands as a complex adaptive system, developed with three colleagues (Janssen *et al.*, 2000).

An Agent-based Model of NSW Rangelands as a Complex System

Ecology of an integrated (biophysical–social–economic) rangeland system

In a series of papers, Perrings and Walker (1995, 1997, 1999) developed an optimal control approach to managing a particular rangeland enterprise. Implicit in this is, of course, the notion that the state of the rangeland can be optimized, or rather that there is an optimal sequence of control activities that will maximize the manager's objective function. From a manager's perspective it is entirely reasonable to attempt to maximize some welfare function over a defined period of time, and knowing what sorts of decisions would yield the optimal outcome under the particular sets of ecological and economic circumstances will clearly be beneficial. However, at a regional scale it is informative to consider the linked dynamics of the whole, integrated system – the biophysical rangeland, the managers and government regulators. Considered in this way, while the individual managers may attempt to maximize their own welfare there is no goal or optimization in the overall dynamics of the system. The various players – grasses, trees, livestock, managers and government regulators are all, in their own ways, attempting to maximize their own welfares. The outcome depends on the rules of the game. We know something about the biophysical rules governing the dynamics of the first three of these, but we are naïve about the behavioural responses of the last two – responses to the biophysical system and to each other. There is no final, equilibrium state in such a system, and very likely no optimal state that best satisfies all the players. Rather, the system is best considered as a continually changing complex system with multiple possible trajectories, some of which may preclude the system from moving into others. The value of attempting such analyses is to learn which combinations of policy rules, manager behavioural and cognitive abilities (which can also be thought of as rules), and ecological and economic conditions result in the system remaining on desired or acceptable trajectories. Clearly, what is important in such a concept of rangeland systems is a knowledge of the rules.

A preliminary attempt to develop an adaptive agent model of a rangeland system in this way (Janssen *et al.*, 2000) highlights the importance of the rules, including the cognitive processes that are attributed to individual managers. Figure 20.1 portrays the structure of the model used.

The model is based on the Western Division of NSW, Australia and includes both the mulga and chenopod land systems. The ecological model includes the dynamics of grass, shrubs and sheep numbers under the influence of rainfall, sheep grazing and fire. It includes an

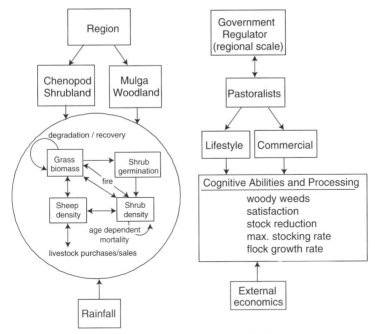

Fig. 20.1.   Structure of an agent-based model of the NSW rangeland system.

age-dependent mortality model for the shrub dynamics that allows for extended periods of shrub domination with reduced grass production. It also includes the 'degradation' of the range under very low cover, and exhibits a slow recovery from such a condition under reduced grazing and good rains, i.e. there is a hysteresis in the pattern of degradation and recovery. It calculates wool production and keeps track of the sales and purchases of livestock.

The socio-economic component allows for different cognitive abilities among pastoralists, resulting in a range of different decision points for each of a number of management decisions, as follows:

- the threshold value of young woody weeds above which the rangeland is burned (only applies to mulga woodland);
- maximum stocking rate relative to grass biomass;
- proportional reduction in livestock in a drought year;
- threshold value for sheep flock growth rate below which destocking occurs;
- degree of satisfaction per unit of consumption.

The socio-economic model includes equations for the pastoralists' utility functions (as indicated above, how satisfied they are per unit of consumption), net income, their financial budgets, the expected flock size, the expected amount of grass biomass, the actual stocking rate, and a rule for renewal of pastoralists to replace those who become bankrupt.

The evolution of the rangeland system was studied under different policy and institutional regimes that affected the behaviour and learning of pastoralists, and hence the state of the ecological system. Three types of regulator policies were used – free market, stabilization (which involved government subsidy during drought) and conservation (control of stocking rates to ensure good ecological condition). Each policy had an associated set of rules and each of 100 pastoralists was randomly assigned a set of parameter values relating to the decision points, described earlier. This set of parameter values in effect describes the pastoralist's 'mental model' of how the rangeland system works under fire and grazing. Together, these rules and the pastoralists' mental models govern the behaviour of the pastoralists. Pastoralists survive or become extinct according to their accumulated profits or debts and new pastoralists with new cognitive abilities replace those that become extinct.

Two results from this model are worth mentioning here. First, after a 200-year run, using real rainfall and wool price data as drivers, the properties under the free market policy (which involved virtually no constraints on sheep stocking rates, regardless of rangeland condition) were in poorer ecological condition than under either of the other two policies and a significantly higher number of pastoralists had become extinct (gone bankrupt). However, taking the average farmer that survived at the end of 200 years (the 'evolved' average set of pastoralist cognitive abilities) and entering such a farmer into another 200-year run, under each policy type, resulted in a property that was in better ecological condition, and with a higher economic return, than was achieved by the equivalent pastoralists who 'evolved' under either of the other two policies. The free market policy allowed the population of farmers to 'learn' through survival of the fittest. In the model, learning was achieved through extinction of farmers with inappropriate management responses to the ecosystem – such as allowing sheep to reduce grass cover too much before selling some of them. The cost of learning, however, was high, in terms of both damage to the rangeland and the social cost of having high numbers of bankrupt farmers.

The second noteworthy point from this preliminary modelling effort concerns the difference in the results from the adaptive agent-based model with the results of an optimal control model that was run using the same ecological model. In the latter, the managers are assumed to have perfect knowledge of future rainfall and wool prices and of the ecosystem responses to management actions. The optimal control model yielded a 30% higher profit than the best adaptive agent model and (since it was designed to do so) also avoided any long-lasting ecological damage. The 30% reduction in profit under the best of the adaptive agent outcomes, plus some ecological damage, can be considered as the cost of 'learning'.

These results are presented only as an indication of the differences between an optimal control approach and an adaptive agent approach to examining policies for rangeland management. Considerably more work is needed, particularly on determining the appropriate cognitive abilities for farmers and on the rules that govern the changes in policy, before the model can be usefully applied. In particular, attention should be paid to how rules are made and changed, and how to model this process – i.e. the rules for making rules. The most valuable use of such models will likely be in exploring which policy combinations allow learning by managers without causing unacceptably high damage to ecosystems, or unacceptably high social costs. These are the two necessary ingredients for sustainability.

References

Atkins, D., Hunt, L.P., Holm, A.McR., Burside, D.G. and Fitzgerald, D.R. (1999) Land-use values in the goldfields of Western Australia and their use in regional resource use planning. In: Eldridge, D. and Freudenberger, D. (eds) *People of the Rangelands. Building the Future. Proceedings of the VI International Rangeland Congress*. VI International Rangeland Congress Inc., Townsville, Australia, pp. 1010–1011.

Bartlett, E.T., van Tassell, L.W. and Mitchell, J.E. (1999) The future use of grazed forages in the United States. In: Eldridge, D. and Freudenberger, D. (eds) *People of the Rangelands. Building the Future. Proceedings of the VI International Rangeland Congress*. VI International Rangeland Congress Inc., Townsville, Australia, pp. 1008–1009.

Bousquet, F., D'Aquino, P., Rouchier, J., Requier-Desjardins, M., Bah, A., Canal, R. and le Page, C. (1999) Rangeland herb and herder mobility in dry intertropical zones: multi-agent systems and adaptation. In: Eldridge, D. and Freudenberger, D. (eds) *People of the Rangelands. Building the Future. Proceedings of the VI International Rangeland Congress*. VI International Rangeland Congress Inc., Townsville, Australia, pp. 831–836.

Fabricius, C. (1999) Evaluating Eden: who are the winners and losers in community wildlife management? In: Eldridge, D. and Freudenberger, D. (eds) *People of the Rangelands. Building the Future. Proceedings of the VI International Rangeland Congress*. VI International Rangeland Congress Inc., Townsville, Australia, pp. 615–623.

Fitzhardinge, G. (1999) People and landscape: growing together or growing apart. In: Eldridge, D. and Freudenberger, D. (eds) *People of the Rangelands. Building the Future. Proceedings of the VI International Rangeland Congress*. VI International Rangeland Congress Inc., Townsville, Australia, pp. 56–61.

Foran, B. and Howden, M. (1999) Nine global drivers of rangeland change. In: Eldridge, D. and Freudenberger, D. (eds) *People of the Rangelands. Building the*

Future. Proceedings of the VI International Rangeland Congress. VI International Rangeland Congress Inc., Townsville, Australia, pp. 7–13.

Janssen, M., Walker, B.H., Langridge, J. and Abel, N. (2000) An adaptive agent model for analysing co-evolution of management and policies in a complex rangeland system. *Ecological Modelling* 131, 249–268.

Lusigi, W. and Acquay, H. (1999) Challenges facing pastoralists in the drylands of sub-Saharan Africa. In: Eldridge, D. and Freudenberger, D. (eds) *People of the Rangelands. Building the Future. Proceedings of the VI International Rangeland Congress.* VI International Rangeland Congress Inc., Townsville, Australia, pp. 83–84.

Lynam, T.J.P. and Dangerfield, J.M. (1999) Fractured, incomplete and practically useless: scientific models of southern African rangeland systems. In: Eldridge, D. and Freudenberger, D. (eds) *People of the Rangelands. Building the Future. Proceedings of the VI International Rangeland Congress.* VI International Rangeland Congress Inc., Townsville, Australia, pp. 825–831.

McCarthy, N. and Swallow, B. (1999) Managing the natural capital of African rangelands: rights, risks and responses. In: Eldridge, D. and Freudenberger, D. (eds) *People of the Rangelands. Building the Future. Proceedings of the VI International Rangeland Congress.* VI International Rangeland Congress Inc., Townsville, Australia, pp. 1002–1007.

Mitchell, J.E., Joyce, L.A. and Bryant, L.D. (1999) Applicability of Montreal process criteria and indicators to rangelands. In: Eldridge, D. and Freudenberger, D. (eds) *People of the Rangelands. Building the Future. Proceedings of the VI International Rangeland Congress.* VI International Rangeland Congress Inc., Townsville, Australia, pp. 183–185.

O'Connor, T. (1999) Community change in rangelands – towards improving our understanding. In: Eldridge, D. and Freudenberger, D. (eds) *People of the Rangelands. Building the Future. Proceedings of the VI International Rangeland Congress.* VI International Rangeland Congress Inc., Townsville, Australia, pp. 203–208.

Perrings, C. and Walker, B. (1995) Biodiversity loss and the economics of discontinuous change in semiarid rangelands. In: Perrings, C., Maler, K.-G., Folke, C., Holling, C.S. and Jansson, B.-O. (eds) *Biodiversity Loss.* Cambridge University Press, New York, pp. 190–210.

Perrings, C. and Walker, B. (1997) Biodiversity, resilience and the control of ecological–economic systems: the case of fire-driven rangelands. *Ecological Economics* 22, 73–83.

Perrings, C. and Walker, B. (1999) Optimal biodiversity conservation in rangelands. In: Eldridge, D. and Freudenberger, D. (eds) *People of the Rangelands. Building the Future. Proceedings of the VI International Rangeland Congress.* VI International Rangeland Congress Inc., Townsville, Australia, pp. 993–1002.

Scholes, R.G. (1999) Does theory make a difference?. In: Eldridge, D. and Freudenberger, D. (eds) *People of the Rangelands. Building the Future. Proceedings of the VI International Rangeland Congress.* VI International Rangeland Congress Inc., Townsville, Australia, pp. 31–34.

Walker, B.H. (1995) Having or eating the rangeland cake: a developed world perspective on future options. In: West, N.E. (ed.) *Rangelands in a Sustainable Biosphere. Proceedings of the Fifth International Rangeland Congress.* Salt Lake City, Utah, pp. 22–28.

Building the Future: a Human Development Perspective

C. Dean Freudenberger

21

Introduction

How can rangeland people be more adequately empowered to improve their quality of life and livelihood? My plenary address, given at the close of the last session of the Congress, represented my response as an ethicist to the contributions of the Congress through the plenary addresses, workshops, and the published two-volume proceedings (Eldridge and Freudenberger, 1999). The following reflections are also based upon my previous experiences with similar consultations and work in the applied fields of agricultural and rural community development in more than twenty nations. As one who has been involved with rangeland projects in Sahelian West Africa following the droughts of the mid and late 1960s, I am no stranger to the challenges faced by rangeland people.

From the perspective of personal observations of the United Nations World Food Conference held in Rome in 1974 and of the United Nations Conference on Desertification in Nairobi in 1977, I conclude that the VI International Rangeland Congress was outstanding. One of the principal reasons for this success was its stated theme: 'People and rangelands: building the future.' This theme encouraged papers and discussions that recognized the need for greater understanding of the two dynamic elements involved in building the future: people and their resources. These two elements are inextricable. As simple a concept as this is, this focus represents a significant leap forward in scientific deliberations about the so-called development process. The Congress struggled with the question of how rangeland science can involve rangeland people in a common effort to build a more sustainable future. This concern was examined within the fluid context of expanding human populations, the post-colonial phenomenon of the globalization of market forces and future shocks of rangeland change (Foran and Howden, 1999). Given this context and a focus on how rangeland science might contribute to improving the welfare of rangeland people on a sustainable basis, one could sense that a new approach was being formulated. My summary statement is a way to support this conclusion by identifying key insights that emerged during the Congress that can contribute to the empowerment of rangeland people to improve their lives and their various enterprises. The Congress participants represented the expectations and needs of some of the world's most marginalized and voiceless people. This in itself is an outstanding contribution to the processes of international social and economic development.

On certain occasions, the utilization of a metaphor is useful for an analysis or a summary of an observation or experience. Given the theme of this Congress about building the future, I shall reference the processes and materials involved with the construction of a building. I recognize the limitation of such an approach, yet the metaphor provides a framework for identifying those insights that emerged during the several Congress sessions that directly contribute to the empowerment of rangeland peoples.

Requisites

Need

Without a sense of need, a building would not have a functional purpose. This Congress addressed the awesome need for reversing deteriorating social, environmental and natural resource trends that have unfolded across almost all of the earth's semi-arid landscapes that are referenced as rangelands. These landscapes directly support the livelihoods of nearly 30% of the world's human population. Involved with the future of these lands are historically unprecedented issues such as human and livestock population growth, species extinction, economic and social inequity, poverty, the dynamics of new post-colonial economic and industrial organizations and new market realities. The lack of creative public policy for providing essential infrastructures for the development of more promising futures is a further illustration of the need for a new building.

Vision

To build, one must have a vision about the design of the building so that it can serve the intended need. In other words, what type of a building is required? One of the highlights of the VI International Rangeland Congress was the continual reference to the vision of 'sustainability'. The goal, as well as the unwritten guideline for building the future of rangelands and rangeland people, is that of sustainability. The term 'sustainability', as referenced by the participants of the Congress, assumed the biological aspects of the regeneration of rangelands. A general understanding prevailed that the term is an ambiguous one that requires continual definition. Today, there is growing recognition that with sufficient skill, and economic and political power, a destructive system can be sustained for a relatively long period of time. During the sessions, one observed that the meaning attached to the visionary concept of sustainability is a reflection of the United Nations Commission on the Environment's definition of 'meeting our needs without compromising the ability of future generations to meet their own needs' (World Commission on Environment and Development, 1987). Incorporated within this definition is the concept of the welfare of future generations. This understanding assumes that future generations of both human and non-human life make moral claims upon our present generation's concept of freedom and responsibility. Involved in this concept of sustainability, widely explored at the United Nations Earth Summit meeting in Rio de Janeiro in 1992, are the ethics of trans-generational and inter-species justice. As the Congress recognized the problem of deteriorating rangelands, it searched for ways to reverse the trend and work for more sustainable, or regenerative, futures. The vision represents not only a challenge of great magnitude, but also a radical understanding of trans-generational and inter-species justice. In other words, if there is no justice in social and economic spheres, sustaining the health of landscapes with their full biodiversity is problematic. If there is justice within the contemporary world, but it is not sustainable over time, then the system is self-defeating.

The vision of sustainability, interfaced with the sense of need for structuring a new future, contributed to serious deliberations. Radical thoughts were expressed. The possibility for actualizing normative ideas about building a more sustainable future was questioned. However, it should be recognized that history often reveals that today's radical ideas have a way of becoming tomorrow's conventionalities. In the opening session of the Congress, Tim Flannery suggested that the responsibility of the Congress was to 'toss out mad ideas'. In various ways, the Congress accepted this challenge. The delegates continually talked about building sustainable futures. This vision resulted in serious discussion, particularly within the context of the problem that the series of rangeland congresses have been addressing during the past 24 years.

Design

In addition to identifying the need for building and articulating a vision of sustainable futures, the questions must be asked: What is needed to meet building codes that are enacted to guarantee structural integrity for public safety? How are the architectural designs engineered to meet the codes? From my ethicist's viewpoint, identifying structural requirements is one of the foremost tasks of the scientist. A building envisioned to meet a particular need cannot be created without the design engineers.

Several relevant elements of rangeland science are involved in response to the relatively new challenge of working towards sustainable futures in collaboration with rangeland stakeholders within a rapidly and radically changed world (Freudenberger and Freudenberger, 1994). Relevant rangeland science addresses urgent needs facing us now as well as for the future. It helps to recognize not only the connection between rangeland health and the stability of planetary health, but also the consequent need for new codes of behaviour in relation to these fragile ecosystems. Relevant rangeland science provides the basic research for designing sustainable human relations with the earth. Relevant rangeland science predicts long-range consequences. Scientists need to take the risk of predictability. Many critical issues of our time, such as the impact of ozone shield erosion and the accumulation of greenhouse gases are only partly understood. However, in spite of uncertainty with reference to these matters, it may be best to lessen the human impact upon these systems now rather than wait for conclusive evidence. For example, the Union of Concerned Scientists, with many Nobel laureate members, demonstrates responsibility for prediction. Relevant rangeland science research articulates and tests principles of regenerative practices for sustainable development. This type of work challenges prevailing anthropocentric orientations and values, and points to the need for radical change in human self-understanding about its relationships and responsibilities with the land. Relevant rangeland science is informed by multidisciplinary conversations. E.O. Wilson in *Consilience: the Unity of Knowledge* (1998) demonstrates the critical importance of interdisciplinary approaches as a means of unifying the knowledge of the many scientific and liberal arts fields in order to see the big picture or the forest, instead of simply the individual trees. Relevant rangeland science contributes to 21st century ethics as it suggests the need for new human value constructs such as the value of the health of the land and people, and advocates the creation of infrastructures for enabling the practice of stewardship. Relevant rangeland science informs and recommends social, economic and industrial policy for sustainable development. Thus, it is engaged in the process of public policy formulation. The new challenge for rangeland scientists is to contribute more actively to public understanding for effective policy development. The final paragraph of the session preface to 'Future shocks to people and rangelands' of the Congress proceedings well summarizes the challenge:

> The final judges of our usefulness are not range management professions but the public, politicians, international bureaucrats and the collateral disciplines that are consumers of our research results. They ultimately set our agenda, sideline us or make us central to the determination of public policy, and allocate or withhold the money needed for continued research.
>
> (Behnke and Howden, 1999)

The majority of papers of the two volumes of proceedings and the workshops were devoted to the management of plant communities and grazing practices, the integration of the management of land and water resources, understanding about plant functional types, the dynamics of plant invasions and the necessity for the maintenance of rangeland diversity. The diversity of these fields points to what is involved in the design of sustainable rangelands and the public policy educational task. The integrity of natural systems must be respected and preserved if building the future of rangelands and their peoples is to be a possibility rather than simply a vision.

Building Blocks

In addition to a conviction regarding the need for a building and a thoughtful and well-informed vision for how the structure will address the need, a building must have building blocks. I have personally made thousands of both kiln-fired and earth-rammed bricks and concrete blocks. Thinking about building blocks is second nature to me. As the Congress proceeded with its deliberations, I identified five categories of building blocks that were not available over the past 35 years. These new building blocks suggest why we may be at least slightly optimistic about the prospects of actualizing a sustainable future for rangelands and rangeland peoples. They suggest how rangeland people might be better empowered to improve the quality of their lives and livelihoods.

Humility

Repeatedly, participants acknowledged past failures in policy, programme design and programme execution. The many changes in contemporary social, political and economic structures point to the difficulties in planning and application of project concepts. The magnitude of the problem of landscape degradation is a sobering element in designing ways to address the many complex and interrelated challenges of building more promising futures. Discussions involving the leaders of major international and national development organizations, as well as non-governmental organizations, pointed to the realization that past approaches and research methodologies do not provide models for designing new initiatives. Past failures were acknowledged. Thus, there was hesitancy or tentativeness in the process of conceptualizing how to take the next steps in rangeland development. These admissions were frustrating, yet creatively so. The heady, self-assured and confident attitudes that prevailed during the past several decades are subsiding, resulting in more thoughtful approaches for the design and implementation of research agendas, projects and programmes. The challenge of functioning within the context of this ambiguity resulted from a recognition that at present, there are not many answers. I evaluate this new insight as a sign of maturity. Insights have been gained by past mistakes and false starts. A new sense of humility about responding to the challenges that lie ahead can be considered a positive contribution in formulating new initiatives for building the future.

Networks

An essential building block involves collaboration in research and its practical application. Case studies illustrated the value of the networking process, particularly for rangeland scientists who are so isolated from each other due to distance and difficulties in communication. Old acquaintances were renewed. New friendships for collaboration were made. The Congress contributed greatly to the expansion of the network of people dedicated to the development of more sustainable futures for rangeland and rangeland people. This building block needs to be understood as a most essential component. Expanding the network should be encouraged.

Perspectives

The stage for the deliberations of the Congress was set within the first plenary address by Tim Flannery. He reminded delegates of the time spans and evolutionary processes that cover hundreds of millions of years that were involved in the formation of today's rangelands. He illustrated the diversity of the history and structure of rangelands by contrasting the differences between North American and Australian landscapes. A plenary panel was devoted to offering perspectives from China, Russia, Peru and Australia. Consequently, the Congress was cautioned about generalities in rangeland issue analysis and prescriptions for addressing the problem of deterioration on a global scale. Flannery's address also dramatized the complex and fragile nature of the earth's landscapes, and so reminded all the participants of the profound responsibilities that rangeland people have accepted. These perspectives challenge traditions of reductionist methodologies in the pursuit of knowledge and encourage greater study of the ecology of the ecosystem within which one is working. Overview perspectives about time and the evolution of biospheric processes contributed to broadening economic analysis and accounting methodologies that forced the question of how to incorporate environmental and resource impact concerns into economic measurements of external costs. Broadened perspectives about rangelands, the functions that they perform in biospheric processes, and of the impact of human enterprises upon rangelands contribute to an understanding of the critical importance of rangeland science in pursuit of global sustainability.

Participatory engagement

One of the outstanding building blocks identified during the Congress deliberation was that of the necessity of community collaboration, or participation of all the stakeholders, for building the future. The theme of this Congress guaranteed discussion of this topic. The workshops and the Congress papers on 'Rangelands: people, perceptions and perspectives' (Mills and Blench, 1999) and 'Indigenous people in rangelands' (Griffin and Fourmile, 1999) emphasized the importance of incorporating cultural values, aspirations, feelings and indigenous wisdom into the process of searching for ways to reverse social and environmental degradation. There emerged in these discussions the insistence that project planning, implementation, evaluation and, if necessary, reformulation, involve all stakeholders. An ever-growing appreciation of indigenous wisdom about rangeland management and cultural patterns of relationships with both human and natural communities was apparent. Serious discussion unfolded about the rationale of pre-colonial patterns of rangeland utilization. The new forces of population growth and global market dynamics point to the limitations of earlier patterns of occupation as models for determining systems for post-colonial rangeland management. These insights and perspectives are essential elements, or building blocks, for a new structure. Interfacing science with indigenous wisdom is of obvious importance. However, from the colonial era to the present phenomenon of globalization, this obvious connection has too often been ignored.

Diversity

The importance of maximizing both biological and human cultural diversity was widely emphasized during the Congress discussions. Today, as never before, biodiversity is clearly understood as essential for landscape stability. This recognition was not prevalent 50 years ago. Given the maturing appreciation of indigenous wisdom and cultural patterns of rangeland management, the Congress discussions underscored the importance of maximizing cultural diversity. This is a critically important insight that is now playing a determinant role in research for the building of a more sustainable and therefore promising future.

A personal challenge

Five categories of building blocks that go into the construction of a new building have been identified. With great seriousness the Congress laboured the attitude of humility in taking next steps, the importance of expanding the networks of communication, broadening perspectives on its science, gaining clarity about the importance of participatory engagement, and the necessity of maximizing and preserving biological diversity. As obvious as these issues may be in our time, one must be reminded that these orientations did not determine the use of range-lands in the industrialized nations nor influence so-called interna-tional development strategies following European decolonization of the 1960s. These emerging insights are of vast significance. These building blocks must be incorporated into all scientific pursuits. The intent for suggesting the use of the metaphor of constructing a building and in identifying what has been evaluated as emerging insights of the Congress is to suggest that this process of identifying categories of building blocks be continued. The prospects for the future depend on sharing insights about these issues.

Mortar

Without mortar, building blocks cannot be joined together. This Congress did not talk about mortar. In the case of the construction of the building that the Congress has envisioned, mortar can be translated to the word 'motivation'. What is it that will empower us to pursue the vision of sustainable futures? Some will say that we are motivated by enlightened self-interest. In other words, if we do not respond to the need to reverse the process of landscape and cultural deterioration, then in time we will all become victims of exploitation, resource loss and environmental degradation. Some say that we had better 'wake up', come to our senses and build for a sustainable future, but is this type of motivation enough to empower us over a long period of time to cope with such obstacles as cynicism, ill-formed public policy, corruption, loneliness, and of how to apply our knowledge and wisdom in meaningful ways?

One suggestion that can be proposed is that the mortar needed to hold the building blocks together for constructing an enduring future is a sense of care. Caring is what makes us human. Caring is the most effective way, if not the only way, for expressing a sense of gratitude for life and our moment of existence within the evolving history of the living planet. Caring about building the future of rangelands and its people is the mortar that will assure the integrity and endurance of the structure that we have committed ourselves to build.

References

Behnke, R. and Howden, M. (1999) Session preface: future shocks to people and rangelands. In: Eldridge, D. and Freudenberger, D. (eds) *People of the Rangelands. Building the Future. Proceedings of the VI International Rangeland Congress*. VI International Rangeland Congress Inc., Townsville, Australia, p. 6.

Eldridge, D. and Freudenberger, D. (eds) (1999) *People and Rangelands Building the Future: Proceedings of the VI International Rangeland Congress*. VI International Rangeland Congress Inc., Townsville, Australia, 1064 pp.

Foran, B. and Howden, S.M. (1999) Nine global drivers of rangeland change. In: Eldridge, D. and Freudenberger, D. (eds) *People of the Rangelands. Building the Future. Proceedings of the VI International Rangeland Congress*. VI International Rangeland Congress Inc., Townsville, Australia, pp. 7–21.

Freudenberger, D.O. and Freudenberger, C.D. (1994) Good relationships: ethical and ecological perspectives on rangeland management. *The Rangeland Journal* 16, 321–332.

Griffin, G. and Fourmile, H. (1999) Indigenous people in rangelands. In: Eldridge, D. and Freudenberger, D. (eds) *People of the Rangelands. Building the Future. Proceedings of the VI International Rangeland Congress*. VI International Rangeland Congress Inc., Townsville, Australia, pp. 385–415.

Mills, D. and Blench, R. (1999) Rangelands: people, perceptions, perspectives. In:
    Eldridge, D. and Freudenberger, D. (eds) *People of the Rangelands. Building the
    Future. Proceedings of the VI International Rangeland Congress*. VI International
    Rangeland Congress Inc., Townsville, Australia, pp. 49–87.
Wilson, E.O. (1998) *Consilience: the Unity of Knowledge*. Random House,
    New York, 367 pp.
World Commission on Environment and Development (1987) *Our Common Future*.
    Oxford University Press, Oxford, 41 pp.

Synthesis: New Visions and Prospects for Rangelands

22

Kenneth C. Hodgkinson, Ronald B. Hacker and Anthony C. Grice

Introduction

The scientific programme of the VI International Rangeland Congress (IRC) held in 1999 in Townsville, Australia, attempted to break new ground. The mandate to move outside a traditional framework was given by the Australian Rangeland Society, which hosted the Congress. The organizing committee determined that the focus should be on the needs and aspirations of rangeland people. The aim was to promote new visions and prospects for the future among the international community of scientists and other people under the theme: 'People and rangelands: building the future.'

Earlier congresses recognized the importance of the social and economic systems within rangelands but rarely were their treatments oriented towards more than one sector, the pastoral industry. Presentations and discussions on the dynamics of ecological systems, and the influence of grazing upon them, dominated earlier congresses. Neglect of social and economic systems was perceived by the organizing committee (and some keynote and concluding speakers at the V IRC held in Salt Lake City in 1995) to be now limiting the application of ecological knowledge. The scientific programme was organized by two of us (Tony Grice and Ken Hodgkinson) with the intent of shifting the emphasis towards people issues.

The preceding chapters arose from the conduct of the scientific sessions. These were written by convenors, based on issues highlighted by keynote speakers, posters that were linked to each session and

©CAB *International* 2002. *Global Rangelands: Progress and Prospects* (eds A.C. Grice and K.C. Hodgkinson)

general discussion at the end. They are succinct summaries of the state of knowledge in most areas of rangeland science.

Our task has been to derive from these summaries a synthesis that will serve both as a benchmark for progress in rangeland science and the underpinning of visions for future prospects. We emphasize the plural because no single vision is appropriate; the rangelands are too diverse in people, cultures, landscapes and land use. We accept that the Australian environment, and the institutions we have worked in, have inevitably shaped our perspective and will bias our synthesis. To minimize this we have used as our information source only the summaries provided by session convenors. These convenors were selected to provide country and gender balance and their contributions have been fundamental to our task. Their work before, during and after the Congress is gratefully acknowledged. We hope that by avoiding the use of references we have prepared a readable synthesis.

The Initial Vision

Fifty years ago in the USA, modern rangeland management was born and seen to be the science and art of obtaining maximum livestock production from rangeland consistent with conservation of land resources. People in the rapidly developing range science profession, both inside and outside the USA, subsequently focused on the livestock production industry for it was the dominant land use and was facing a number of challenges from within and without. The greatest challenge was to find the means of achieving sustainability and maintaining natural resources for future generations. The vision appears to have been that well conducted science would achieve sustainable livestock production from rangelands and, by implication, viable rangeland businesses and communities.

Systems for monitoring the health of rangelands (range condition) and, to a lesser extent, grazing management dominated the thinking of scientists, extension workers and public institutions and became a worldwide focus. This concentrated the research on soil–plant–livestock–climate relationships. Furthermore, the science was largely conducted independently of pastoral business operations or subsistence production systems. Many scientists and most rangeland people, especially graziers and indigenous pastoralists, saw range science as an outside activity, external to their day-to-day concerns.

The vision was neither clearly articulated, nor shared, and increasingly it proved inadequate. It was too simplistic. There are important industries in rangelands other than pastoralism and they may all interact significantly. The sustainability of a family pastoral business, for example, may depend on a member having work in a nearby mining operation, or on the running of an adjunct tourism enterprise. The interconnected biophysical, economic and social systems are each vitally important and if one fails, sustainability may be jeopardized. Furthermore, the transfer of Western scientific knowledge to developing countries, without appropriate modification, has not been successful. Now a rethinking is required which will more comprehensively accommodate the realities of rangelands and the people who use and live in them.

New Visions and Prospects

Communities and individuals are increasingly developing their own visions in response to local and regional issues. Issues are usually complex, generally operate over multiple spatial scales and are continuously changing. The visions therefore need to be adaptive so that lessons can be learnt from successes and failures and from changing internal and external environments. Issues cannot be universal; each region or area has its own special people and resource characteristics. At the moment, people and governments are experiencing difficulty in thinking about complex issues in a holistic manner. However, there are examples around the world of communities developing their own plans for action. Rangeland people will shape their own futures; it is likely that science will be only a small part of this process.

The obvious task for rangeland professionals is to produce management tools in collaboration with rangeland people. These tools will enable their communities to develop more inclusive, comprehensive frameworks for analysing rangeland issues and planning appropriate responses. The frameworks must encompass the social, economic and biophysical systems and identify the pathways for implementation of the decisions made by local institutions and groups. If rangeland science can address these issues it will contribute materially to diminishing the effects of marginalization, developing adequate lifestyles and living infrastructures, reducing poverty and improving education while maintaining the traditional role of developing resilient systems, which maintain resource function.

Progress and Prospects for Rangeland People

The new directions we have briefly sketched above offer many opportunities for productive partnerships between professionals and the people of rangeland communities. Realizing these opportunities will require an understanding of the current condition and needs of rangeland people and the institutions that serve both people and the needs of biophysical resources. This Congress has provided an opportunity to look at progress and assess prospects.

Indigenous people

Rangelands are home to many groups of indigenous peoples around the globe. The survival of their cultures is intimately linked with the range-lands; rangeland resources provide their spirituality, culture in the broadest sense and sustenance. They are an intimate part of finely tuned natural systems and are therefore very vulnerable to changes in their local and larger environment.

European colonization in India, North America, Australia and Africa has had a devastating impact on these peoples. This alienation and displacement of indigenous peoples over recent centuries has created massive poverty, inequality and the degradation of traditional lifestyles and cultures of these people. The means of restoring these things are not readily apparent. In some areas, the emergence and spread of human and indigenous rights movements has brought about some restoration but inequalities largely remain. Furthermore, the problems are growing. In the so-called developed countries, growing human populations, a scarcity of land and demand for commercial products are placing enormous pressure on the natural resources and on the indigenous people themselves.

In the developing countries, indigenous peoples face even more challenges. The population here will more than double over the next 100 years. Economic globalization will present an increasing threat to survival of many indigenous societies. Expenditure on rangeland development projects for indigenous peoples has declined since the early 1980s. Today the emphasis in rangeland development projects has shifted from increasing primary production and technology transfer to projects that involve communities and their traditional ecological knowledge for sufficient duration to foster lasting benefits for indigenous peoples. Hopefully the new strategies emerging to incorporate traditional ecological knowledge into resource management will provide mechanisms for indigenous people to benefit from the more effective use of their knowledge.

One-third of the earth's land surfaces are rangelands and there is a need to acknowledge the valuable ecological services provided by the rangelands to indigenous and non-indigenous peoples alike. To achieve understanding of these non-production or ecosystem service roles, and to develop socially acceptable plans to maintain them, all groups of people should be represented. *Participation by representatives of indigenous peoples and all stakeholders will be a prerequisite for success in the implementation of policies and institutional change in rangeland regions.*

Grazier people

For graziers in the developed countries, recent decades have seen a continuing decline in the economic and political importance of the rangeland livestock industries. Economically, rangeland production systems can make only a minor contribution to world trade and graziers will continue to be driven by it with no ability to control the outcomes. A continuing and perhaps accelerating decline in the terms of trade appears likely.

Rangeland communities throughout developed countries operate in an environment in which the previous alignment between their goals and aspirations and those of the wider community can no longer be taken for granted. In this situation of conflict over land management objectives, the critical issues concern the extent to which, on the one hand, scientific understanding is sufficient to underpin land management for alternative uses and, on the other, the policy and regulatory framework promotes socially and politically acceptable outcomes.

Development of cost-sharing mechanisms that balance public and private interests is now a key to the achievement of land use that is both socially and ecologically sustainable. The importance of this area of policy development is heightened by the two fundamental characteristics of rangelands – variable climate and long response times – that favour short-term decision-making by graziers in a market-based economy. In these circumstances, conflict between private and public interests in land management is most likely to occur in periods of climatic stress. Priority should be given to *development of policy instruments that are linked to measures of landscape condition and aimed at maintenance of rangeland productivity under these conditions – these remain a challenge for graziers and institutions in developed countries.*

Nowhere is the need for cost-sharing mechanisms more evident than in the quest for biodiversity conservation. Given that existing (and likely future) nature reserves fail to adequately conserve a sufficient cross-section of environments, the preservation of rangeland biodiversity will require off-reserve conservation to supplement the more formal reserve system. Implementation of such an approach will require resource assessment and planning at regional scales and a policy framework that encourages graziers to balance their immediate self-interest against broader, often intangible, public and ecological values. The provision of financial incentives for graziers, by either government or private interest groups, will be an important part of this framework. Local decision-making will also be pivotal in off-reserve conservation. It is an important role for government and private interest groups to provide incentives to help local graziers bear the cost of conserving biodiversity.

Regional resource assessment and planning which must underpin reconciliation of diverging public and private interests in turn requires appropriate consultation frameworks in which graziers are adequately represented. These frameworks should incorporate the capacity to analyse issues in social, economic and biophysical terms and identify pathways for implementation. Local ownership of decisions is pivotal to the success of off-reserve conservation.

Divergent public and private interests in rangelands represent both an opportunity and a threat for graziers. The implementation of incentives or other cost-sharing arrangements to support the achievement of public benefits would provide an alternative income stream with potential to reduce inter-annual fluctuations in income. Such developments, together with increasing use of rangelands for recreational and other non-consumptive uses, would also serve to increase the dialogue and connectivity between city and rangeland people. On the other hand, closer scrutiny by the urban public and the need to accommodate alternative value systems may, for some, only exacerbate a sense of being marginalized.

There appear to be few new technologies likely to improve management practices of graziers in developed countries in the short to medium term, although some are emerging. Some existing technologies are not yet widely adopted. Rangeland monitoring methodology is well developed, for example, but its potential to contribute to business decision-making remains largely unrealized. Even when implemented by graziers, monitoring is usually seen to be more about demonstrating environmental credentials than about enterprise management. *Graziers need to work with scientists to explore how to use monitoring methodology and other tools, more effectively.*

The use of landscape function analysis to help identify management objectives or to monitor longer-term management impacts has prospects for widespread use but is limited by the lack of developed models for interpretation of quantitative data. The growing sophistication of ecosystem models, including the spatial dimension, together with the remarkable advances in computing power and Internet communications do provide great opportunities for application to day-to-day management but, with the possible exception of the US, widespread realization of this potential seems unlikely in the short term.

While we have advocated an increased emphasis on policy development, we acknowledge that understanding of some aspects of the biophysical system requires further refinement. Weed infestation remains a major concern to graziers throughout the developed countries. The area of exotic weed-infested rangeland is increasing because there are few economically viable solutions to plant invasions. The extent to which weed invasions can be halted by grazing management practices alone is problematical but these will often be the only control measures available. They must be combined with a strategic approach, prioritizing weeds on the basis of their actual impact, and involving risk assessment at the ecosystem (rather than the species) level.

Urban people

The people that live in cities are mostly ignorant about what goes on in rangelands, but this varies from country to country. In Australia, most people live in coastal cities distant from the rangelands whereas in the USA, China and other countries, there are large cities within rangelands. Generally, city people use the products from rangelands without thinking about their origin or their own 'ecological footprint'. Urban growth continues at a significant rate and the prospects are that *rangelands will continue to be distanced from the real affairs of the majority of urban people and their political decisions.*

There are no easy ways of addressing this marginalization and rangeland people will need to continue mustering all the political clout within their means, to address the specific issues that emerge from this imbalance, neglect and ignorance. Communities working together with each other and institutions to achieve their agreed goals will go a long way towards achieving adequate support and understanding from urban dwellers and city-based institutions.

There are good prospects that the growing environmental crises will increase the dialogue and connectivity between urban and rangeland people. Conservation of biodiversity, supply of clean water, meeting of international obligations, availability of 'clean and green' meat, availability of game and feral animals for hunting, and wide open spaces for recreation and spiritual renewal are some of the issues of, and products from, the rangelands that will keep rural and urban people connected. It is increasingly apparent too that in developed countries cost sharing to balance private need and public good is a key issue and this alone will foster better communication.

Institutional people

Governments and their departments are constantly setting up and modifying policy, implementing programmes to address perceived problems, and providing services to people and industries within their jurisdiction. The reasons for these activities are to influence people in the way they utilize natural resources, in the way they relate to each other (within and across state/prefecture/other and national boundaries) and relate to their environment. Government intervention in rangelands usually involves rangeland-specific regulations as well as general policy on a wide range of issues. Public (and private) institutions are quite influential on the management of natural resources through policies and regulations but prospects are that range-land people will continue to be ambivalent or negative about specific policies and regulations.

Policy-makers need information and expert advice to support their work, and the information/advice they need may be drawn from a broad spectrum of sources and disciplines. They want information and advice that have some capacity to predict the likely outcome of proposed, and alternative, policy settings.

In our view the Congress highlighted four major issues for institutions.

1. There is the importance of assessing or auditing the status of rangeland resources at business, regional and national scales. This is needed to assess the basic long-term capacity of the resource to produce goods and services, and to determine management effectiveness. Given the importance of assessment it follows that auditing at the regional, national or even international level should be based on statistically sound approaches designed for each specific purpose. The attributes measured must be adequately defined so that others may form their own opinion of the relevance of the attribute to the question being asked.

2. Recent new understanding about grouping plants into functional types may prove to be most useful in rangeland planning and manage-ment at regional scales. It now seems possible to provide institutions with a way of indicating vegetation change associated with manage-ment activities and environmental changes as well as indicating the sustainability of rangeland ecosystems. Identifying appropriate plant functional types will provide an essential underpinning of the knowledge relevant to national and international planning and policy-making.

3. It is now apparent that existing nature reserves in rangelands do not adequately conserve a sufficient cross-section of natural environments. It is therefore essential to foster off-reserve conservation to supplement the reserve systems across rangelands. Local decision-making is clearly pivotal in off-reserve conservation but there is also an important role for government and non-government interest groups to provide incen-tives to help local landowners bear the cost of conserving biodiversity in both developed and developing countries.

4. The people of rangelands inevitably have distinctive views developed by imitation of others, especially parents, and life's experiences that influence the way they manage land and other resources. Even where there are multiple uses, most rangeland people have a single-use view of rangelands. This 'unimodal' approach is inadequate for institutions responsible for the long-term welfare of the community as a whole because they cannot focus solely on the needs of any one group. The difficulty in pursuing a multi-modal approach is that governments do so in the face of conflict between differing interest groups, each attempting to ensure that their principal benefit is maintained. Inevitably, compromise between competing uses and objectives has to be reached because none has a pre-ordained right to dominate. Participation by representatives of stakeholders is a prerequisite for success in implementing policies and institutional changes in rangeland regions. *Prospects for participation by stakeholders in developing policies and changing institutions are now good as effective approaches are available.*

Scientific people

Some question whether scientists should be numbered among rangeland people. Nevertheless, scientists are drawn to the rangelands, want to serve rangeland people and, through research, contribute to their social and economic well-being and the sustainability of rangeland resources. The uniqueness of rangeland ecosystems and the people fascinate them. However, there is a divide, sometimes very wide, between what scientists produce and the expectations of other rangeland people. Rangeland people may support scientific research in principle but the products are often ignored and they rely only on their personal local knowledge. This congress provided clear messages to scientists that it is urgent that this gulf be bridged.

Part of the problem is that scientific and local knowledge clearly develop in different ways. Science creates models to structure information and evaluates the models by experimentation. The organizing, validating and transporting of knowledge is conducted in a well-developed impersonal manner through space and time and allows science to 'claim' universality. In contrast, local knowledge develops from observation and 'imitation'. Indigenous peoples have local knowledge systems of long standing. Graziers are of comparatively recent origin and their local knowledge is different. Just as each local 'community' comprises a mix of often diverse and disparate social groups and cultures, so local knowledge systems contain varied and contested ways of knowing and managing the same local environments, all of which may be different from those of 'outsiders' such as conservationists, government officials and scientists. Indigenous and other local knowledge is also not static – it grows over time and in response to new stimuli. Local knowledge is also context-dependent, like science, and failure to appreciate this leads to poor uptake of technological solutions in rural development.

The challenge ahead is to build more effective bridges between scientists and other rangeland people. Bridges do exist but these should be strengthened and many new ones built. New strategies are emerging throughout the world's rangelands that will help incorporate traditional ecological knowledge into scientific thinking about resource management. Knowledge from practical experience of grazier people should also be used as a platform for innovation and operational scale experiments that involve graziers and scientists working together in partnership. The congress recognized that these bridges will be most effective in promoting change if they are built on the adaptive capacity of people.

Respect for all forms of local knowledge is important for social cohesion, regardless of the perceived utility of that knowledge in promoting production or ecological goals such as biodiversity conservation. Rangeland scientists must therefore include different human values and perceptions in their understanding of resources and resource use. Scientists can contribute to the development and implementation of the strategies developed by rangeland communities, but in many cases this will require a substantial change in their *modus operandi* towards approaches that integrate biophysical, cultural and economic goals.

Information or insight has to be communicated to, and accepted by, other people before it becomes part of a system of knowledge. Scientists in general do this poorly. Trust needs to be strengthened so that scientists and other rangeland people alike accept exchanged information as valid knowledge. New ways need to be found for integrating science with local knowledge in ways that promote sustainability.

A number of knowledge areas were identified at the Congress as being in need of more development. One area is desertification that affects millions of rangeland people. Progress with managing desertification has been slow and was identified as a continuing problem for many indigenous and grazier people. It must remain an area for scientific research. Rangeland scientists must better understand the social and economic systems for these now contain the 'bottlenecks'. We know how landscapes lose function and productive potential when overgrazed. This knowledge, coupled with the development and application of on-ground and remotely sensed methods for determining change in landscape function, now place the necessary technical tools for locating where desertification is occurring within the grasp of all rangeland people.

However, there are major deficiencies in our knowledge of soil processes. One is the biology of soil processes. The web of organisms involved in the decomposition of organic matter is not understood in sufficient detail. The survival of key organisms during desertification and the consequences if they fail to survive are poorly known. Furthermore, there is a need to integrate knowledge of soils across scales and to know more about the alternative pathways and 'critical thresholds' in soil processes. Growth in study of the fine-scale relationships of within-soil processes would assist the management of both desertification and biodiversity.

Another area of inadequate knowledge is how to scale-up knowledge to address vegetation dynamics at larger scales. A central problem encountered when scaling vegetation responses to regional levels is the limited ability to quantify and interpret complex floristic responses involving a large number of individual species. It is now possible to group plant species in terms of both their responses to disturbance and their effects on ecosystem processes. Prospects for global comparisons now depend on development of standardized approaches for identifying the general patterns of vegetation responses. Rangeland scientists also need to know to what extent species may replace each other without a loss of system function. Use of plant functional types for indicating likely vegetation change and the responses of rangeland ecosystems to disturbance deserves further development.

An important and growing aspect of rangeland management is the control of exotic and native weeds. In rangelands these are typically shrubs and grasses of low palatability. They affect the profitability of pastoral businesses and influence biodiversity. The area of weed-infested rangeland is increasing because there are few economically viable solutions to plant invasions of the rangelands. Occasionally we make inroads into rangeland weeds through less expensive technologies such as fire and biological control, but globally, the weeds are winning the war. Clearly weeds are the consequence of human colonization and it is rangeland people that must manage weeds or learn to live with them. Weed management in rangeland ecosystems requires research to develop more strategic approaches with a focus on risk assessment of ecosystems rather than species.

Increasingly, graziers are required to keep an eye on both production and conservation goals in their pastoral businesses. Conservation of biodiversity, although a 'fuzzy' concept, is often an undervalued by-product of rangeland grazing. Conservation ecology and its relationship with social and economic systems is clearly a growth area. Scientists are increasingly required to develop appropriate methods to assess biodiversity; point-in-time and point-in-space methods are inadequate. The use of surrogates for biodiversity levels, such as landscape function analysis, should be further evaluated. Scientist and manager alike must better understand the benefits of ecological services and contributions to the resilience of rangelands provided by biodiversity.

The exchange of knowledge between scientists and rangeland managers takes place in many ways but the future must include computer modelling for simulation. Scientists already have a number of models that achieve acceptable predictions of biomass change at paddock, property and regional scales. The impact of climate and market variation on commercial production of meat and fibre can now easily be evaluated to address 'what if' type questions. Models, termed agent-based models, are now being developed to include the complexity and interactions between social, economic and biophysical systems. As rangeland analysts and modellers come to grips with integrating the human dimension of rangelands into their ecological models and understanding (or vice versa), they face the severe problem of the limited compatibility, both theoretical and practical, between most current economic models and ecological models. When these and other problems are overcome, the developments will be useful to a wide range of people who are responsible for grappling with the behaviour of complex systems in the rangelands and elsewhere.

Making the existing and future models useful to a section of range-land people managing at different spatial scales remains a big problem. For example, how do scientists present the important concept of uncertainty to different audiences in ways that have meaning to them? What is required is a more inclusive, comprehensive framework for analysing rangeland issues which incorporates social, economic and ecological aspects and which provides pathways for implementation of the decisions made by local institutions. Despite the many challenges that face modellers there are remarkable opportunities to be seized in Internet technology, computing power and conceptualization. These opportunities will make the next decade a very exciting time for modellers and users of models. Communication between scientist and manager need no longer be a top-down or one-way process. *The question of how best to capitalize on these developments in modelling, and involve rangeland people so that they have a sense of ownership, requires urgent consideration by scientist and user alike.*

Understanding the dynamics of rangeland vegetation, and the influence of fire, grazing and drought, has been a popular activity of scientists. It is a well-researched area. However, the myriads of studies have not provided a high degree of predictive capacity. More data is not the immediate answer. More understanding and new perspectives are urgently needed. Scientists should be endeavouring to construct coherent conceptual frameworks for predicting the consequences of factor interactions. This should be done at a range of scales so that emergent properties at higher scales are modelled and understood. Biological variability and complex interactions between organisms should be included in model validation experiments, rather than avoided, even if it means less clear mechanistic insight. Natural resource administrators and science programme managers should therefore support the construction of coherent conceptual frameworks for understanding how factors interact to influence the dynamics of vegetation across relevant scales.

We perceive that grazing management research will diminish. There is now good predictive understanding of how individual plants respond to grazing, how grazing regimes influence the dynamics of plant populations and when landscapes, with their assemblages of plant and animal communities, respond to grazing pressure. Total grazing pressure, rather than grazing system, appears to be the major factor in grazing management. The processes can be modelled. It is now time to shift some, but not all, resources from this area. There is still a need to better understand the effects of drought and how grazing can be most appropriately managed before and during these times. This research is in progress.

The major lesson for scientists from this congress was to *build into their teams, those with social and economic system skills*. The new science approach unfolding now in rangelands is the management of complex adaptive systems. This shift requires a different combination of research skills and outlooks from those that were prevalent in the last 50 years of the old millennium.

Concluding Remarks

The VI International Rangeland Congress provided yet again, an important forum for exchange of ideas and experiences and for renewing and making new friendships with people around the world. The extent to which the scientific programme was able to significantly influence and shift the 'mind-sets' of participants is difficult to assess. Hopefully, participating scientists and members of public and private institutions especially, will more fully appreciate the needs of rangeland people and the importance of searching for and achieving sustainable social, biophysical and economic systems in their rangelands.

Rangelands of the world will continue to be marginalized from the global economy, most urban people and national governments. However, they remain the home of millions of people, and contain and supply much that is important to non-rangeland people. Future meetings of the International Rangeland Congress and its sister organization, the International Grassland Congress, will continue to be important for supporting rangeland people. The challenge for the future of both congresses will be to more comprehensively discuss the social and economic systems as well as the biophysical system in the context of the many and varied visions that rangeland people have for their futures.

\

Index

Page numbers in **bold** refer to figures and tables

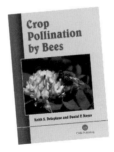
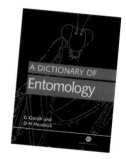
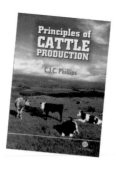